Butterfli
on British and Irish offshore islands:
Ecology and Biogeography

Butterflies on British and Irish offshore islands:

Ecology and Biogeography

Roger L. H. Dennis
The Manchester Museum,
Manchester University, Oxford Road,
Manchester M13 9PL

&

Tim G. Shreeve
School of Biological & Molecular Sciences,
Oxford Brookes University,
Oxford OX3 0BP

GEM PUBLISHING COMPANY, WALLINGFORD

Gem Publishing Company
Brightwell cum Sotwell
Wallingford, Oxfordshire OX10 0QD

First published December 1996

British Library Cataloguing in Publication Data.
A catalogue record for this book is available from the British Library.

ISBN 0 906802 06 7

Typeset by Gem Publishing Company, Wallingford, Oxfordshire.
Printed in England by Swindon Press Ltd, Swindon, Wiltshire.

For
Tom Dunn
and the late
Ian Lorimer

CONTENTS

FIGURES

TABLES

PREFACE

The publication of this work on the butterflies of Britain's and Ireland's offshore islands has two objectives. The first is to encourage more thorough observations on the butterfly species found on islands around the British and Irish coastlines, by those who have the good fortune to be able to visit them. To this end we provide a simple list of the species that have been observed on 219 islands. In the process of recording species on islands, we would encourage observers to obtain as much information as possible to determine the resident or vagrant status of species. The reasons for making such distinctions are discussed in the text and advice is given on how to prepare for recording and what to look for during visits to islands (Appendix 2). As islands have long been productive settings for ecological and allied research, the list with its accompanying bibliography should also enable those involved in such research to decide where work may usefully be conducted on particular species. Our second and main objective, then, is to encourage ecological and evolutionary research into island faunas, particularly with regard to questions involving the conservation of biological diversity in Britain and Europe. In this publication, we contribute to this programme by investigating the ecological significance of species' records for the British and Irish islands.

The publication falls into two sections. The first presents the results of analyses for a selection of 73 islands. Together with a standard appraisal of island species' richness, this includes an assessment of faunal differences among islands and of geographical affinities among species. Brief reviews are also provided on the ecology of butterflies on the offshore islands and their geographical variation, emphasis being placed on determining the status of species on the islands. Plant names follow Clapham, Tutin & Moore (1989). The results confirm the relatively limited influence of island area and isolation in accounting for the numbers of species on the islands. The British butterfly species are well able to migrate to nearby offshore islands. Evidence is accumulating that they are also suited to colonize islands and to survive on them for extended periods of time. Species' richness on islands and the incidence of species on islands is shown to be most closely correlated to the species' richness and the incidence of species at faunal sources on nearby islands and on the adjacent mainland of Britain, France and Ireland. Not surprisingly, this is related to the range and distribution of species in Britain. However, covarying with and underlying species' richness and the incidence of species on islands and at faunal sources, as well as their ranges and distributions on the mainland areas, are a number of ecological parameters. They, it is argued, influence migration, colonization and survival. From these results we present a straightforward ecological interpretation of the occurrence of butterflies on British islands.

Although the findings would suggest a further shift away from expectations based on historical viewpoints, some historical signals are emitted by the island data. Caution is advised in regarding historical factors and influences on island faunas as of little significance. One problem that has become evident is that present day patterns can effectively mimic those produced by historical processes. Regarding biogeographical data, the difficulty remains of discriminating been factors in historical and ecological time.

The second section documents the species found on islands. The records are given in the form of a simple list for each island following the order in the check list. The list is linked to an extensive bibliography including references up to 1 August 1996. For a complete bibliography on the entomology of smaller British islands, the reader is referred to Smith & Smith (1983). To avoid much repetition and to reduce the bulk of the species' list for islands, references to literature on island records are numbered according to their order in the bibliography and documented for groups of islands. A number of the records presented here are previously unpublished. Some are on file at the ITE Biological Records Centre, Abbots Ripton, Huntingdon, but many have been received as personal communications in response to requests for records made in the entomological journals; these are referred to sources enumerated in a separate list. There is insufficient space to give details of these records here; they are to be stored with the British Butterfly Conservation Society's Recorder as part of *Butterfly Net* and the *Millennium Atlas Project* (see Appendix 2).

The order and names of taxa in the check list follow Emmet & Heath (1989) and Dennis (1992). Species frequently recorded for the British and Irish mainlands (i.e., native species; common and infrequent immigrants) are registered in the main check list and their current status indicated. More unusual records (i.e., rare immigrants and vagrants, deliberate introductions and accidental introductions) are listed in Appendix 1. Two date classes are provided (pre- and post-1960) to give some indication of the antiquity of many of the records. No data are given on the breeding status of butterflies on the islands; this is often unclear and such status varies with time. As analyses are conducted that make use of geographic variation in species, data on infraspecific variation (subspecies and other geographical forms) are documented in the list, but no comment is made on the validity of the nomenclature for subspecies. In most cases, further work and revision is required.

ACKNOWLEDGEMENTS

We extend our grateful thanks to recorders who have kindly sent us their observations of butterflies on the British and Irish offshore islands (see section II.3). We are particularly indebted to those who have contributed a large number of records: Paul T. Harding of the ITE, Biological Records Centre at Monks Wood Experimental Station, Abbots Ripton, Huntingdon, Cambridge; Tim A. Lavery (Scientific Director) and Karen Cronin (Research Assistant) of the Irish Lepidoptera Records Database, Country Watch, c/o Regional Technical College, Tralee, Co. Kerry, Ireland; Ian Rippey of Portadown, Co. Armagh, Northern Ireland; Dr Jack Gibson (The Scottish Natural History Library, Kilbarchan, Renfrewshire) and Dr George Thomson in Scotland. Dr Gibson very kindly gave much of his time to scan the literature for us in the Scottish Natural History Library. We are very grateful to Vicky Seegar for her hospitality during the visits of one of us to Hilbre Island, to the Hilbre Bird Observatory for records from the island, to Barry T. Shaw for making us aware of data for this Cheshire island, and to Ian Rutherford for the loan of island literature.

This short book has benefited substantially from the careful editorial work of Gerry Tremewan. We owe our greatest debt to him for his painstaking correction of numerous errors and for ensuring that we have conveyed some points more clearly than we might have done.

The work is dedicated to two colleagues, the late Ian S. Lorimer of Orkney and to Tom C. Dunn of Chester-le-Street, Co. Durham, who have made valuable contributions to the knowledge of entomology on the British islands and who have always given generously their advice and assistance.

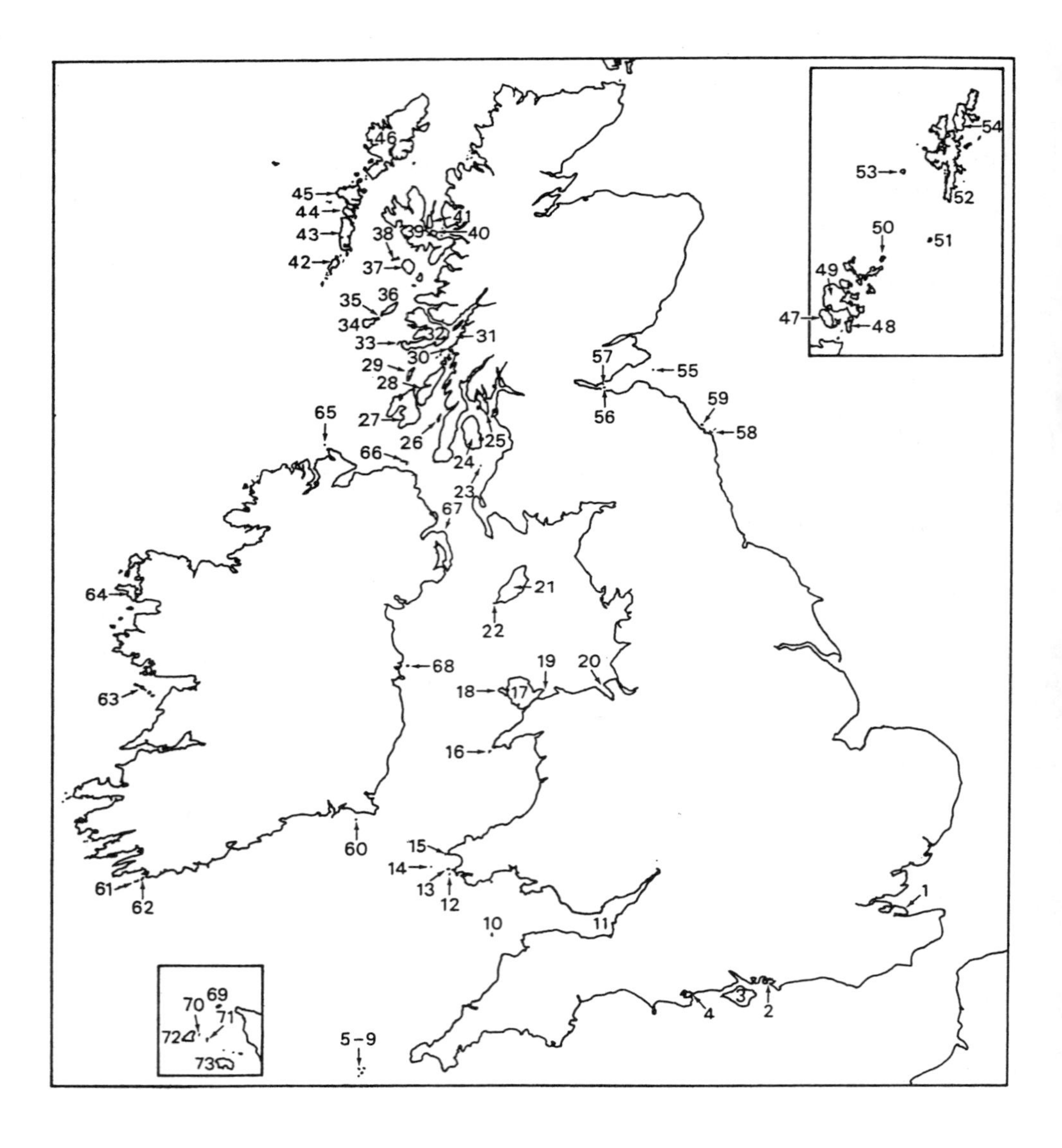

Fig. 1. The location of the 73 British and Irish islands used for biogeographical analyses. See Table 1 for names.

I. BIOGEOGRAPHY OF BUTTERFLIES ON BRITISH AND IRISH OFFSHORE ISLANDS

1. ECOLOGY OF BUTTERFLIES ON ISLANDS

Enquiries into island faunas often treat species as equivalent elements. However, the status of species may differ substantially. Ideally, for comparative research, data should be collected on breeding status, hostplants, habitats, population size, colonization events, extinctions and introductions. Often such details are unavailable, but the information that exists on these topics is very revealing. The present section has two objectives. First, we hope to encourage more thorough observations; much valuable information can be obtained even during short visits to islands (see Appendix 2). Second, details that exist on butterfly ecology are of great importance in developing an understanding of the composition of island faunas, their biogeography and, ultimately, advance their conservation. In this section, we demonstrate how the records of butterflies on British islands can be given ecological interpretation. As such, we hope to encourage research on islands where this may increase our knowledge of butterfly ecology and evolution.

A. Breeding records, hostplants and habitats

Observations of butterfly species on islands tend to be of adults. Usually, there is little attempt to search for early stages or for other evidence of breeding and island residence. The quality of records varies enormously. Particularly good records were made by the late Professor Heslop Harrison, who went to great lengths to determine the status of species in the Scottish Western Isles whilst carrying out his research on the botany of the islands. Some islands have had resident entomologists (e.g., Canna: Campbell, 1975a) or have had wardens with a particular interest in entomology (e.g., Rhum: Wormell, 1982; Farnes: Walton, J., pers. comm.) and contribute to the Butterfly Monitoring Scheme (e.g., Skokholm; Skomer; Pollard & Yates, 1993). But, the records for many islands emanate from visitors, only few of whom have the detailed recording of insects and other wildlife as a primary objective. There is some indication that this may be changing, with an increasing effort being made to determine the status of species on islands (e.g., Shetland islands: Harper, 1974; Harvey, Riddiford & Riddiford, 1992; Pennington, 1993, pers. comm.), often in conjunction with research being done on island botany (e.g., Farnes: Hirons, 1994; Stringer, G., pers. comm.). It is well to appreciate what is required. Observations of mating (e.g., *Pieris brassicae* on

Grassholm: Heron, 1956), or of egglaying, are insufficient proof for residence on islands. Survival from the egg stage can be very low (Warren, 1992); indeed, butterfly species are occasionally forced to use unusual hostplants when their normal hostplants are absent or scarce (e.g., *Polyommatus icarus* on Tresco, egglaying on *Ulex europaeus*; Sutton, pers. comm.) and survival on these unusual hosts is rarely recorded. More substantial records are those of larvae and pupae, when correctly identified and found in numbers, and of newly emerging, teneral adults (e.g., *Pieris napi* on the Eye Peninsula, Lewis: Heslop Harrison, 1956b).

Sins of commission occur as well as those of omission. Not all remarks on the absence of breeding are justified. The early stages have been sought at the wrong times for some species (e.g., search for *Anthocharis cardamines* eggs on Rathlin Island in late June and early July: Rippey, pers. comm.) and it is clear that some observers lack knowledge of hostplants and habitats. For example, it was once thought that the swarms of *Pieris napi* seen in the Inner and Outer Hebrides were the result of mass migration and not of breeding on the islands. However, Heslop Harrison (1947d, 1950c, 1955b) provided ample evidence that *Pieris napi* breeds extensively there on a number of hostplants. This issue may have arisen owing to the butterfly favouring small rather than large hostplants for egglaying (Dennis, 1985). Often, pertinent details to support comments about the status of species are absent (e.g., Hilbre: Craggs, 1982). Consequently, island faunas cannot at present be compared on the basis of so-called breeding records and we have not entered such detail in the lists for islands. It can be easily sought in the accompanying references.

Superficially, many islands would appear to provide very poor conditions for the survival of butterfly species. An example may be Clare Island, off Mayo, Eire, which Kane (1912) described as lacking both dune and forest and dominated by vast tracts of wet bog and moorland yet with twelve recorded species. However, even apparently inhospitable islands in northern Britain may have several species breeding on them from time to time. For example, twelve species have been recorded from Handa, off north-west Scotland, at least seven of which have resident populations. These observations were made by Derek Hulme during 176 visits to the island between 1964 and 1993. Although specific hostplants may be recorded as absent from islands (e.g., *Urtica dioica* for *Aglais urticae* on North Rona: Dannreuther, 1939), negative records are not always reliable. Hostplants and suitable habitats can be tucked away in inaccessible locations, as in the case of the *Boloria selene* colony confined to precipitous cliffs on Sanday (Campbell, 1970). Some islands are little more than sea-washed wave-cut platforms but still have butterflies breeding on them. Heisker, a mere reef 9.8 m above sea level to the south-west of Canna in the Inner Hebrides, has breeding records for *Pieris napi*, *Polyommatus icarus* and *Aglais urticae* (Heslop Harrison, 1955b; Campbell, 1970); *A. urticae* was found to be using *Urtica urens* near the lighthouse (Heslop Harrison, 1938c). On some other islands alternative hostplants must be used although it is not always obvious what these may be (e.g., *Gonepteryx rhamni* in the Channel Islands: Long, 1970). Trees are notably scarce on most British islands and so too are butterfly species dependent on them, but there are exceptions. For example, the only records of *Nymphalis polychloros* in the Isles of Scilly have been for Tresco, corresponding to the presence of elm in the abbey gardens (Richardson & Mere, 1958). Particularly noteworthy is the record of *Quercusia quercus* from Colonsay (Dunn, 1965). This observation came from the native woodland remnant at Coille Mhor in the north east of the island, the oaks dwarfed and wind-blasted into a *krummholz* condition. A final cautionary tale on negative records comes from observations on moths. The frequent presence of tree-feeding moths belies the absence of woodland on some islands (e.g., Canna). It is highly unlikely that sea barriers are obstacles to moth dispersal but it is also interesting that the relevant tree species have subsequently been found, deep in ravines and on cliff sides, often prostrate to rock surfaces (Heslop Harrison, 1947a).

Surveys carried out by competent entomologists reveal that not only are hostplants often available for butterflies on islands (e.g., Bardsey: Morgan, 1969; Askew, 1974), but also a variety of habitats are shown to exist for many butterfly species and different hostplants for them in these habitats. It is worth illustrating this point in some detail from the valuable work done by Heslop Harrison for the Scottish Western Isles; it demonstrates the importance of carrying out thorough surveys on islands, even over what may initially appear to be unrewarding ground. It also demonstrates the value of applying different techniques, such as sweeping for larvae or simply searching for larvae directly by looking for feeding damage, to find butterflies that are not found flying during surveys (e.g., *Callophrys rubi* from *Empetrum nigrum* on Scalpay: Heslop Harrison, 1937a). In summarizing his many observations on the butterflies of the Outer Hebrides, Heslop Harrison (1950c) noted the variety of habitats used by them. Among sand dunes, machair, cliff tops, grass banks, lower moor slopes, moorland tops, boggy depressions, lakesides, runnels and drainage lodes, the machair ranked as the most important habitat for butterfly species. Nevertheless, *Polyommatus icarus* was not only found on sand dunes and the machair, where *Lotus corniculatus* would be expected to abound, but also on cliff tops and open moorland where it used the same hostplant. *Argynnis aglaja* also frequented open moorland, using *Viola riviniana* growing under heather, as well as marshy places, dunes and machair where it used *Viola palustris* and *V. curtisii* in addition to *V. riviniana*. *Pieris napi* was found to breed on *Cardamine* species on the moorlands, on *Nasturtium officinale* and *N. microphyllum* on marshy ground in the vicinity of the machair, and on *Cochlearia officinalis* on cliffs and rocks. Summer visitors are not exempt from utilizing different resources in different habitats. On Rhum, *Cynthia cardui* bred on *Cirsium arvense* near the coast but on *Carlina vulgaris* up in the glens (Heslop Harrison, 1944). Different species were also found to use alternative habitats and different hostplants on separate islands (Heslop Harrison, 1950c, 1955a).

Data on other aspects of butterfly biology are more scarce. Yet, surveys carried out on islands can be particularly valuable for extending knowledge on the ecology of butterflies, particularly the relationship between migration, voltinism, herbivory, parasitism and population dynamics. For example, observations from the Orkney Islands and the Outer Hebrides have distinguished resident from migrant populations of *Pieris brassicae* (Lorimer, 1983; Heslop Harrison, 1938a, c); at times, the population structure of this butterfly in the Outer Hebrides is known to be very complicated, with as many as four or five different cohorts of individuals existing in different stages simultaneously, some resident, some migrant, some parasitized and others parasitoid-free (Heslop Harrison, 1942b). *P. brassicae* has been shown to exploit crops other than cabbages, 'millions' of individuals derived from a single field of turnips on Rhum in 1945 (Heslop Harrison, 1946b), as well as native herbs occurring among crops (e.g., *Sinapis arvensis* among cereals on Canna: Campbell, 1970). Parasitization of such pest species as *P. brassicae* is supposed to occur seldom on islands according to some entomologists (Campbell, 1970), but there is evidence for occasional heavy parasitization in even isolated northern islands (Heslop Harrison, 1942b). Islands, because of their isolation, clearly provide useful experimental systems for studying changes in animal populations. However, continuous records over long periods of time are needed. This can be illustrated by the differences of opinion as to the status of *Inachis io* in the Hebrides during the first half of this century (Heslop Harrison, 1947b; Campbell, 1969b, 1984). The issue, whether it was vagrant or resident, can now be easily understood in the context of metapopulation dynamics and data on environmental changes (see Harrison, 1991); both opinions would have been correct, but at different times and for different islands.

B. Populations, colonization and extinction

Island biogeography theory relates species' richness on islands to colonization and extinction events. However, observations of colonizations and extinctions are rare. Without data from long-term detailed surveys it is not really possible to know if an organism has previously existed, or continues to persist, in very small numbers. Autecological studies on individual species, which incorporate population estimates, ensure some objectivity but these are typically short-term in nature. Thus, the colonization and extinction process affecting species is still poorly understood. It is generally appreciated that small populations more readily become extinct, and that from large populations there may be more propagules (perhaps even disproportionately more propagules where density is high) – leading to an increase in out-of-habitat movement – which could generate more colonization events on other islands, marine or terrestrial. But, quantitative data for different species are lacking, and it is only possible to speculate as to how island populations respond to environmental changes, mainly from monitoring work done on the mainland (see Pollard & Yates, 1993).

Most work on island butterfly populations has been carried out in the Isles of Scilly on *Maniola jurtina* and *Polyommatus icarus* by Professors Ford and Dowdeswell and their colleagues (Ford, 1964; Dowdeswell, 1981). Their research, extending over 35 years, from 1938, mainly focused on evolutionary processes underlying wing pattern variation. Even so, some interesting data emerged purely on populations, which confirm more casual observations. First, small populations can persist for many years on small islands, as in the case of *M. jurtina* on White Island. Second, extinction does occur, a phenomenon that they were able to confirm from their experimental introductions of *M. jurtina* on Menawethan and Great Innisvouls. Thirdly, substantial changes in population structure, particularly metapopulation geography, can occur within islands in relation to habitat changes, probably best illustrated by their observations on Tean, but also from work on

Great Ganilly and White Island. Finally, perhaps most interesting, they have shown that very large populations can build up on small islands. Prior to 1953, both Tean and St Helens are calculated, using MRR, to have had some 15,000-20,000 *M. jurtina* in any year (i.e., a minimum of 750 to 1250 per hectare). Such numbers may have implications for movements of individuals between islands (see section I.10 and Fig. 11).

When populations have not been assessed by survey, it is probably more difficult to establish whether populations are small or large. Species may appear to be rare simply because the visit has not coincided with the peak of the flight season, that it has been conducted in poor weather or that only a fraction of the island's area or habitats has been investigated. More convincing statements of persistent rarity derive from observations of resident entomologists (e.g., *Pararge aegeria* on Canna: Campbell, 1970; *Polyommatus icarus* on Orkney mainland: Lorimer, 1983; pers. comm.). Even so, it is difficult to translate descriptions of abundance into figures. Massive populations of notable migrants have been observed, both of adults (e.g., *Pieris brassicae* on Hilbre, Cheshire: Blackler, 1940) and larvae (e.g., *Cynthia cardui* on South Uist: Heslop Harrison, 1943e). Obviously very large numbers of resident species have also been recorded (e.g., *Pieris napi* – 'like petals, everywhere' – on Bardsey, Wales, in 1994: McCormick via Hardy, pers. comm.; *Maniola jurtina* on Alderney in 1994, with estimates of more than one million individuals: Moon, pers. comm.). More surprising, perhaps, are the reports of huge populations that occur from time to time among resident species on northern Scottish islands. Many butterflies are typically colonial in northern Britain (e.g., *Pieris napi, Maniola jurtina*: Dennis & Shreeve, pers. obs.) but occasionally they are in sufficient numbers to disperse from colonies to cover whole islands. Thus, 'snowstorms' of *Pieris napi* have been observed on Eigg (Dannreuther, 1936) and Sanday (Campbell, 1952). Vine-Hall (1969) described the numbers of *Erebia aethiops* on the open moors of Mull as being into 'hundreds of thousands', the adults seen flying in woods and along the main street and harbour of Tobermory, the emergence coinciding with 'sweltering hot' weather. Large fluctuations in numbers of species have also been witnessed on islands (e.g., *Inachis io* on Canna: Campbell, 1970), a feature typical of northern Britain (Pollard, Hall & Bibby, 1986). But, with the notable exception of work on *Maniola jurtina* in the Isles of Scilly (Ford, 1964), little attempt has been made to quantify these changes (but see Thomas, Moss & Pollard, 1994).

Colonizations and extinctions are as difficult to substantiate on islands as they are for mainland habitats (Hardy, Hind & Dennis, 1993). Their determination depends largely on consistency in the monitoring of species and the quality of records on breeding. Even so, in defining colonization and extinction, because records ultimately lack certainty and because surveys only provide estimates from samples, it is difficult to avoid the use of arbitrary periods of absence, prior to the event for colonization and after the event for extinction (see Pollard & Yates, 1993). Owing to the sparseness of observations on islands, many colonization and extinction events must have been missed. For example, it is possible that *Pyronia tithonus* may have colonized Alderney since 1970, since no records exist for it prior to that year (Long, 1970), yet it was plentiful in 1994 (Moon, pers. comm.). The same may be true for *Boloria selene* on Sark, which was to be seen everywhere on the island in May and June 1947, but never seen there before 1939 (Shayer, 1947). In neither case, however, are these records *proof* of recent colonization. What may seem remarkable is that during the last 50 years, when recording has been most intensive, there has been no long-term colonization event (i.e., lasting >10 years) on the large landmass of the British mainland; this, despite the fact that several species have been able to migrate to Britain (e.g., *Papilio machaon*, *Nymphalis antiopa*) and others have been released here (e.g., *Araschnia levana*, *Melitaea didyma*, *Aporia crataegi*, *Iphiclides podalirius*). It may be that the British Isles are marginal for continental species (Dennis, 1977, 1993; Shreeve, Dennis & Pullin, 1996) or that environmental conditions associated with successful migration and colonization have not occurred during this period (see section I.10).

Extinctions are more easily established than colonizations, since, unlike the latter, there is recourse to confirmation by intensive survey at any time in the future. Assuming that there is good evidence that a species existed previously on an island, the only danger is that a subsequent recolonization event can conceal an earlier extinction. Extinctions that have occurred on the British mainland (e.g., *Aporia crataegi*, *Lycaena dispar*, *Cyaniris semiargus*, *Maculinea arion*, *Carterocephalus palaemon* in England) tend to confirm the marginality of the islands for continental species. Extinctions have also been registered for some of the larger islands. Both the Isle of Man and the Isle of Wight may each have lost six species during the period of records (Chalmers-Hunt, 1970; Fearnehough, 1972). The record for extinctions on small islands is extremely poor, even though more extinctions are to expected for them. However, some of the smaller islands are known to have lost species that could well have been part of a resident fauna. For example, *Coenonympha tullia* has probably become extinct on Colonsay and Canna (Dunn, pers. comm.; Campbell, 1975a) and *Plebejus argus* on Guernsey (Long, pers. comm.). Occasionally losses on small islands reflect extinctions over a wider region, as in the case of *Nymphalis polychloros* which is no longer on Tresco, Lundy or Sheppey.

As butterfly colonizations and extinctions may be frequent on small islands, evidence for such events depends more on systematic population surveys. Required, at very least, are resident entomologists making continuous records of breeding and emergence for species. However, records for most small islands are the product of casual visits. There are exceptions. Canna is unique, since continuous records of butterflies were made for it from 1938 to 1975 (Campbell, 1969b, 1970, 1972, 1975a). During this period, two species are known to have become extinct on the island (i.e., *Inachis io*, *Coenonympha tullia*), and a further three have colonized it (i.e., *Boloria selene*, *Pararge aegeria*, *Inachis io*), though some of the species clearly have been involved in a number of extinction and colonization events (e.g., *I. io*). *Boloria selene* appeared in Haligary gully, Canna, for the first time in June 1969 and presumably colonized from Sanday where it was found in 1957. Some islands, of course, are subject to similar systematic landuse changes that have severely reduced butterfly populations on the British mainland, a process affecting even the most northern islands (e.g., dune excavation for construction on Burray, Orkney: Lorimer, pers. comm.).

C. Introductions

Introductions may be classified as intentional or unintentional (see Conservation Committee of Butterfly Conservation, 1995, for definitions of terms allied to introductions). Many unintentional translocations of insects across the British Isles may have occurred in the Holocene, but there can be no record of them. However, it is possible that *Heteropterus morpheus* may have been inadvertently introduced to Jersey from France during the second world war, as a considerable amount of fodder was imported for horses between 1940 and 1945 (Long, 1970). It is a common insect on the French coast opposite to Jersey (Quinette & Lepertel, 1993). Among intentional translocations of butterflies to islands, there have been recent attempts to introduce several species to the the Isle of Man (i.e., *Gonepteryx rhamni, Polygonia c-album, Pararge aegeria*) (Oates & Warren, 1990; Rippey, pers. comm.). For G. *rhamni* and P. *aegeria*, these may represent re-establishments, as there are earlier, if unconfirmed, records (Chalmers-Hunt, 1970). The introduction of G. *rhamni* to the Isle of Man is being accompanied by the planting of the hostplant, *Frangula alnus*. However, as the species is close to the edge of its geographical range in Man, long-term survival will probably depend on growing the hostplant extensively over the island in a variety of habitats. Whilst this effort may demonstrate how a species may persist at the edge of its range, it makes little contribution to island biogeography theory and effort would probably be better directed to conservation of the existing fauna.

The introductions, again probably re-establishments, of *Maniola jurtina* on the small islands (< 0.5 ha) of Menawethan and Great Innisvouls in the Isles of Scilly is documented in detail by Dowdeswell (1981) and Ford (1964). In 1954, 120 and 117 females from St Martin's were liberated on the two islands respectively. The populations hung on precariously in subsequent years; that on Menawethan was abandoned, whilst that on Great Innisvouls was reinforced in 1956 with a further 106 females from St Martin's and monitored off and on until 1964, when only a single male was observed. Why neither experimental population succeeded in establishing itself remained undetermined, though Dowdeswell suggested that the strong winds and the lack of shelter together with predation from a flourishing colony of wrens may have been largely responsible.

The purpose of introductions needs to be carefully considered. It is a simple matter to release large numbers of individuals into the wild without thought for the consequences or without official blessing (e.g., release of 50 *Aglais urticae* from the Italian Dolomites on Lewis in 1995; Hackett, pers. comm.). The difficulties in reconstructing events after introductions are well illustrated by the unsuccessful attempt to obtain information on an alleged transfer of *Lysandra bellargus* and *L. coridon* to Anglesey (Dennis, 1974). The reader is referred to Oates & Warren (1990) for advice but it is not a practice we endorse or encourage.

D. Mistaken records

One of the greatest difficulties in compiling species' lists is the validity of records. Invalid records can be the product of deliberate hoax or fraud or simply result from human error. Thankfully, deliberate fraud is apparently rare, but one needs to be aware of it. There are still those who would collect butterflies and consequently provide a ready market for those who trade in them; a price on an insect, but also occasionally pranks or aberrant notions of importance, can be the reason behind generating false records. A number of examples are cited in the journals for the British mainland and for Ireland. However, there is little reason for such deliberate counterfeit for the smaller British islands. Campbell (1975b) suggests that Heslop Harrison may have been the subject of a prank regarding his *Maculinea arion* record for Rhum, but it seems as if we will never know the circumstances underlying it.

Unintentional mistakes occur more frequently than many atlas recorders would care to admit. There is the difficulty of challenging participants who give so much of their time to surveys. The fact of the matter is that observers do make mistakes, witnessed time and again during conducted walks (Dennis, pers. obs.). Some errors are bound to occur in the simple transcription of records, as in the case of those for *Coenonympha pamphilus* for *C. tullia* in Orkney (Harding, pers. comm.). Mistakes more frequently involve observations in the field. Despite the small size of the British fauna, there is nevertheless potential for confusion among pairs of species even by the most experienced of observers (e.g., *Pieris rapae* and *P. napi*; *Thymelicus sylvestris* and *T. lineola*; *Boloria euphrosyne* and *B. selene*; see Appendix 2). The possibility also exists of racial variation causing confusion. As another reason for the alleged occurrence of *Maculinea arion* on Rhum, Campbell (1975b) suggests that it may have been confused for the large bright *Polyommatus icarus mariscolore*. However, the two are very dissimilar. Nevertheless, it is likely that the alleged records of *Argynnis paphia* and *A. adippe* on Islay were in fact *Argynnis aglaja scotica* (Heslop Harrison, 1941d; Wilks, 1945a; Rippey, pers. comm.). Some moth species may have been mistaken for butterflies, for example, *Odezia atrata* (L.) for *Cupido minimus* and *Euclidia glyphica* (L.) for *Erynnis tages* (see Jeffcoate, 1994). Long (1970) discusses a number of more profound possible mis-identifications in the Channel Islands. These problems underline the value of voucher specimens in the past and of photographs in the present and future.

One type of record that presents a recurring problem is of a single individual of a species previously unrecorded and subsequently unobserved. A number of such records occur for Rhum, again from Heslop Harrison (i.e., for *Erebia aethiops*, *Boloria euphrosyne*, *Eurodryas aurinia*: Heslop Harrison, 1955b) and, without firm evidence for them, it is correct to challenge them (Wormell, 1982). Nevertheless, it is well to appreciate that subsequent absence of a species is not *evidence* that it was never found on an island, nor is the presence of a species evidence that it must always have been resident. Continuous records for islands and gardens and work on metapopulations (e.g., Thomas, Thomas & Warren, 1992) indicate that 'colonizations' and 'extinctions' are frequent; Moreover, long-term observations on populations indicate that both population fluctuations and dispersal can be extremely variable and distinctly periodic (Pollard & Yates, 1993). Thus, records of isolated individuals may represent direct observation of *potential* colonists.

2. ANALYSIS OF ISLAND RECORDS

Analysis of island records involves both the selection of islands and of species. This is not a simple matter as the island records are not the product of a systematic survey based on an appropriate sampling design. The data for each island were not originally collected with any analysis in mind. Moreover, the quality of records for islands clearly varies in a number of important ways: in the number of visits made, their timing and their geographical coverage. Observations have also varied in their thoroughness. Some recorders have searched for early stages, but most have not made any attempt to validate observations as evidence of breeding on islands. Thus, selection of islands for analysis has, of necessity, to be arbitrary. They have been selected on the basis of records for the two non-resident long-distance migrants *Vanessa atalanta* and *Cynthia cardui* and the two resident long-distance migrants *Pieris brassicae* and *P. rapae*. These migrant species have the highest probability of successful sea crossing to islands in large numbers and therefore of being observed on them. As the two pierids are highly apparent, appear at somewhat different times of the year compared to the two nymphalids and have the potential for colonizing inhabited islands with Cruciferae crops (*Brassica* spp.), they have a high probability of being recorded by resident or visiting observers on islands. Therefore, these four species provide some measure of the completeness of an island's list. As of 1 May, 1995, all four species were recorded from seventy-three islands (excluding the British mainland and Ireland; Fig. 1). To provide a comparative 'base line' in multivariate analyses, a further eight units have been added to the file: Ireland, the British mainland divided into four latitudinal belts (<52°N; 52–54°N; 54–56°N; >56°N), northern France (north of 47°N), Belgium and Holland. Treatment of the mainland by region also allows for latitudinal trends in species occurrence to be considered in some analyses (Dennis, 1977). Data for butterflies on the British and Irish mainlands have been extracted from Heath, Thomas & Pollard (1994) and Emmet & Heath (1989), and on the Continent from Bink (1992).

Some selection has also been made of species. Only those species which are habitually resident on the British mainland (and in northern France for analyses including the Continent) have been entered into analyses. Thus, butterflies which may occasionally, but not consistently, survive winter conditions in Britain (e.g., *Colias croceus*, *Vanessa atalanta* and *Cynthia cardui*) are excluded. Again, this decision is necessarily arbitrary, inasmuch as some species (e.g., *Inachis io* and *Aglais urticae*) may have a similar status on many northern Scottish islands. However, for only few islands have long-term continuous records been maintained (e.g., Canna: Campbell, 1970), deterring further distinction of records.

Analyses, including comparative statistics, correlations (Pearson r and Spearman r_s), regression, multivariate ordination and clustering routines (nearest neighbour, complete, unweighted pair-group average) have been carried out using procedures in STATISTICA (1994). A number of useful reference texts on statistical applications in biology are available (e.g., Campbell, 1989; Bailey, 1995). Multivariate techniques used in this work are described in Sneath & Sokal (1973). For those unfamiliar with statistical routines, the following points may be helpful. Means for variables, where reported, are accompanied by standard errors (SE). The standard error is calculated by dividing the standard deviation with the square root of the number of individuals in a sample ($\sqrt{n}$). The standard deviation is a measure of the spread of values around the mean. The standard error describes the bounds within which the population mean is expected to occur about the sample mean with 68% probability; 95% confidence limits are obtained by multiplying the value of the standard error by 1.96. Comparisons have been made using the chi square test (χ^2). Differences are summarized by the test statistic and the so-called degrees of freedom; the latter (indicated by a number in parentheses) measures the size of the test, specifically the number of rows-1 times columns-1 in a contingency table. The probability value gives the likelihood of the test statistic occurring by chance. A probability (*P*) value of less than 0.05 (1/20 of occurring by chance) is regarded formally as a measure of statistical significance. This level of probability for determining significance is also used for correlations and regression parameters. However, it is well to note that in correlation matrices, in which a number of correlations are reported, at least 5% of the correlations may be expected to attain formal significance by chance.

Correlations express the degree of association between two variables. Two kinds of correlations are used in this work. Pearson's zero order correlation coefficient (r) is applied to data on an interval or ratio scale (measures and counts). Spearman's rank correlation coefficient (r_s) is applied to ordinal or ranked data. It is directly equivalent to Pearson's r for higher levels of measurement and can be interpreted in much the same way. The values of correlation coefficients range from –1 (perfect negative association), through zero (no relationship) to +1 (perfect positive association). The amount of variation in one variable statistically accounted for by another is given by the coefficient of determination, simply the square of the correlation coefficient, r^2; when multiplied by 100 it is converted into a percentage. Regression analysis provides a more detailed examination of the relationships between variables. Regression parameters are for the simple linear least squares model; a is the intercept, the value for Y (dependent variable) when X (predictor variable) is zero; b is the slope of the regression line. The higher the value for b, the steeper the slope of the regression line; with higher values for b, each increment for variable Y is influenced by a smaller increment of variable X. For standardized data in which the variates (value for each variable) are given in standard deviation units, the slope is given by beta which is identical to the correlation coefficient. F measures the significance of the relationship. Perhaps the most important aspect of regression analysis is the examination of residuals, the difference between observed values for variable Y and those predicted from values of variable X in the regression analysis. They are illustrated by the scatter of points about the least squares regression line. Large residuals are indicative of a poor fit by the regression equation. In multiple regression analysis, more than one predictor variable is employed to account for the variability in the dependent variable Y. The multiple correlation coefficient and its square are given as R and R^2 respectively. A wide variety of regression models and procedures exist, which can be geared to specific tasks. More is said about this at relevant points in the text. Data have been normalized for Pearson correlations and regression routines so as to meet assumptions for these statistical procedures.

Similarities between species and islands for multivariate routines are Jaccard coefficients (S_J) and affinities are mean Jaccard coefficients ($\bar{S}_J$) (Fig. 2).

		Sample 1	
		present	absent
Sample 2	present	***a***	***b***
	absent	***c***	***d***

a = Number of jointly occurring species in sample 1 and sample 2
b = Number of species present in sample 2 but absent from sample 1
c = Number of species present in sample 1 but absent from sample 2
d = Number of joint absences

a, ***b***, ***c*** and ***d*** are calculated from the total available species pool of all the sampling units.

$$S_J = a / (a+b+c)$$

Fig. 2. The method for calculating the similarity between two samples using Jaccard's similarity coefficient S_J.

One obvious benefit of this coefficient is that it is not weighted by the number of joint absences between the sampling units. Full justification for use of this simple measure of similarity compared to the many available for use is discussed elsewhere (Dennis, Williams & Shreeve, 1991). This coefficient can be used to determine the degree of joint occurrence among species for islands (i.e., islands matrix) or among islands for species (i.e., species matrix). The Jaccard coefficients are processed by multivariate techniques of ordination and clustering. In the search for pattern in the relationships among islands or among species, it is useful to apply a range of ordination and clustering techniques. Agreement between different techniques gives greater credence in the different solutions.

The main ordination method used is non-metric scaling. This produces a 'map' of points, representing islands or species, in two or more dimensions. The points are effectively moved about until their relative positions in the mapped space equates as well as is possible with the ranked differences in their affinities. The programme runs through a number of iterations from a starting configuration, a Guttman-Lingoes initial configuration based on another ordination technique called principal components analysis. Put simply, this seeks to place units (e.g., islands, species) in independent axes, each of which maximally describes the variation among the units. A number of coefficients measure distortion against the original Jaccard coefficients (Kruskal's stress Phi; Guttman's alienation K).

Kruskal's raw stress or Phi:

$$\text{Phi} = \sum(d_{ij} - \text{delta}_{ij})^2.$$

Guttman's coefficient of alienation K:

$$K = [1-(\sum d_{ij} * \text{delta}_{ij})^2 / \sum d_{ij}^2 * \sum \text{delta}_{ij}^2]^{0.5}$$

where
d_{ij} are the observed distances (or dissimilarities);
delta_{ij} are the reproduced distances.

The fidelity of the plot to the original similarity matrix can also be determined from a visual inspection of residuals in a Shepard diagram, in which the inter-unit distances in the plot are matched against the initial matrix of similarities or distances. The plots and coordinates from the final configuration can be compared with those generated from different starting configurations (e.g., the geographical position for islands instead of the Guttman-Lingoes initial configuration). Final configurations can also be compared among different data sets (e.g., island incidence data for species against ecological data for species) or between data and theoretical models (e.g., faunal lists for species generated at random from potential mainland sources).

Cluster analysis encompasses a vast array of different classification algorithms; its objective is the identification of natural clusters. The techniques used here belong to the sequential agglomerative hierarchical non-overlapping (SAHN) genre; they link up pairs of increasingly dissimilar units into a dendrogram. The various techniques differ as to the amalgamation or linkage rules. The three techniques used here represent two solutions at the extremes (single linkage and complete linkage) and a compromise solution (unweighted pair-group method using arithmetic averages, UPGMA). In single linkage, fusion is determined by the two closest objects in clusters, whereas in complete linkage clustering, fusion is dictated by the two most distant objects in clusters. In UPGMA clustering, the distance between two clusters is calculated as the average distance between all pairs of objects in the two different clusters. Single linkage tends to produce a chain of units and is useful for identifying distortion in ordination plots. UPGMA and complete clustering produce tighter clusters; used together they are appropriate for detecting naturally occurring distinct clumps of objects.

3. FACTORS UNDERLYING SPECIES' RICHNESS ON ISLANDS

Three basic factors are generally identified as being of prime importance to the balance of species' numbers on islands: area, isolation and the size of the faunal source (MacArthur & Wilson, 1967; Williamson, 1981). For British butterflies, the theoretical issues involved have been discussed fully elsewhere (Dennis, 1992). Here, only a brief introduction can be given. Equilibrium notions of island biogeography (MacArthur & Wilson, 1967) envisage the number of species on islands as a balance between immigration and extinction. Extinction is primarily modelled on island area. Smaller islands are expected to have fewer species since smaller islands will generally have smaller populations of each species which, stochastically, are subject to higher extinction rates. Immigration is mainly modelled on isolation, the distance to potential sources, which may be other islands or larger land masses. More isolated islands will have fewer species since increasing isolation militates against successful migration, and thus colonization, of potential colonists. The theory conceptualizes a continuous turnover of species, but the maintenance of much the same number of species. However, island area and isolation also affect turnover rates in species. Owing to increased immigration rates, islands closer to land sources have a faster turnover in species than those equal in size but more isolated from shore. Similarly, because of more rapid extinction rates, smaller islands may have faster turnovers than larger islands which are isolated by the same distance from a contributory land source, though immigration rates could be higher for the larger island which presents a larger target. The extremes in turnover are represented by small islands close to shore (fast) and large but isolated islands (slow).

There are, of course, other ways in which island area and isolation may influence species' numbers that do not predict an equilibrium between immigration and extinction. One example is the relationship between island area and habitat heterogeneity; a large island may have a greater variety of habitats than a small island providing opportunities for more species which require different resources. Island geography and resources may also be linked to historical factors which may have influenced colonization in the current interglacial, the Holocene, dating from c. 10 ka BP. These issues, involving the relative significance of equilibrium and non-equilibrium ideas, are explored more fully elsewhere (Dennis, 1992), but see section I.10. Here, more simply, a direct comparison is made between basic factors that may influence butterfly numbers on islands.

Data for variables with which island species' numbers are correlated are given in Table 1; this includes their definition and measurement. Correlations are reported in Table 2. The variables include:

- island area (**A**);
- isolation from the nearest mainland source of France, Britain or Ireland (**I1**);

Table 1. The number of species and geographical data for 73 British and Irish islands (for location of islands see Fig.1).

	ISLAND	S	A	I1	I2	FS1	FS2	LI	L2
1	SHEPPEY	18	8900	0.2	0.2	45	45	51.35	51.34
2	HAYLING	27	1595	0.6	0.6	47	47	50.77	50.88
3	ISLE OF WIGHT	42	38063	2.8	2.8	47	47	50.57	50.77
4	BROWNSEA	30	206	0.4	0.4	48	48	50.69	50.66
5	ST MARTIN'S	14	224	41.0	2.5	36	14	49.96	50.05
6	TRESCO	11	298	46.0	1.5	36	14	49.94	50.05
7	BRYHER	11	125	48.0	0.3	36	12	49.95	50.05
8	ST MARY'S	14	648	44.5	2.5	36	14	49.91	50.05
9	ST AGNES	11	125	50.0	1.8	36	14	49.88	50.05
10	LUNDY	24	423	17.5	17.5	43	43	51.15	51.02
11	STEEPHOLM	13	19	5.0	5.0	46	46	51.33	51.32
12	SKOKHOLM	19	99	3.6	3.6	38	38	51.69	51.71
13	SKOMER	19	293	1.1	1.1	38	38	51.72	51.75
14	GRASSHOLM	3	8	12.8	8.8	38	19	51.76	51.75
15	RAMSEY	13	242	0.8	0.8	38	38	51.84	51.86
16	BARDSEY	13	178	3.4	3.4	33	33	52.73	52.77
17	ANGLESEY	32	71488	0.2	0.2	36	36	53.11	53.22
18	HOLY ISLAND	19	3160	7.5	0.3	36	32	53.23	53.26
19	PUFFIN	8	23	7.0	0.8	37	32	53.31	53.25
20	HILBRE	13	8	1.6	1.6	32	32	53.38	53.38
21	ISLE OF MAN	16	58562	30.0	30.0	24	24	54.05	54.67
22	CALF OF MAN	12	249	57.0	0.5	21	16	54.04	54.64
23	AILSA CRAIG	13	90	13.5	13.5	24	24	55.25	55.20
24	ARRAN	21	43088	16.0	16.0	22	22	55.44	55.70
25	BUTE	14	12352	0.5	0.5	22	22	55.73	55.92
26	GIGHA	11	1413	3.0	3.0	20	20	55.65	55.69
27	ISLAY	18	61887	22.0	0.8	23	18	55.58	55.69
28	JURA	18	60530	4.5	4.5	23	23	55.79	56.14
29	COLONSAY	15	4311	31.5	11.5	23	18	56.03	55.98
30	SEIL	13	1500	0.1	0.1	23	23	56.27	56.31
31	KERRERA	15	1400	0.5	0.5	22	22	56.38	56.40
32	MULL	18	90976	2.0	2.0	22	22	56.27	56.59
33	IONA	10	858	38.5	1.2	22	18	56.31	56.58
34	TIREE	11	7858	36.0	3.1	20	13	56.45	56.71
35	GUNNA	9	66	33.0	0.8	20	13	56.57	56.71
36	COLL	13	7538	14.0	14.0	20	20	56.58	56.71
37	RHUM	14	10717	22.0	13.5	21	18	56.94	56.71
38	CANNA	13	1126	39.0	5.5	21	14	57.04	56.71
39	SKYE	18	161374	0.7	0.7	18	18	57.01	57.03
40	SCALPAY	14	6110	9.0	0.5	17	18	57.27	57.35
41	RAASAY	16	6155	7.0	1.1	17	18	57.33	57.49
42	BARRA	10	9033	77.0	7.0	20	12	56.95	56.71
43	SOUTH UIST	12	31599	75.0	28.0	20	18	57.04	57.49
44	BENBECULA	9	8448	80.0	0.8	17	12	57.00	57.49
45	NORTH UIST	9	34139	74.0	2.5	17	9	57.50	57.49
46	HARRIS-LEWIS	10	205062	41.0	23.5	17	18	57.73	57.86
47	HOY	9	16083	12.5	12.5	13	13	58.77	58.67
48	SOUTH RONALDSAY	8	6090	10.0	5.0	13	9	58.73	58.64
49	MAIN ORKNEY	9	52208	27.5	2.8	13	9	58.87	58.64
50	NORTH RONALDSAY	5	730	86.2	4.0	13	5	59.36	58.64

Table 1 cont.

	ISLAND	S	A	I1	I2	FS1	FS2	LI	L2
51	FAIR ISLE	6	777	139.0	50.0	13	9	59.76	58.64
52	MAIN SHETLAND	5	98480	165.0	40.0	13	6	59.85	58.64
53	FOULA	3	1400	170.0	22.0	13	5	60.15	58.64
54	YELL	4	21900	230.0	2.7	13	5	60.49	58.64
55	MAY	9	49	8.0	8.0	17	17	56.18	56.24
56	CRAMOND	9	13	1.3	1.3	17	17	55.99	55.98
57	INCHCOLM	9	13	1.1	1.1	18	18	56.04	56.05
58	INNER FARNE	9	6	2.4	2.4	18	18	55.61	55.60
59	LINDISFARNE	13	473	1.3	1.3	20	20	55.66	55.64
60	GREAT SALTEE	16	124	6.0	6.0	23	23	52.10	52.17
61	CLEAR	20	639	6.0	1.5	26	20	51.41	51.49
62	SHERKIN	20	497	0.5	0.5	26	26	51.45	51.49
63	INISHMORE	18	1846	12.5	12.5	25	25	53.09	53.22
64	ACHILL	15	9088	0.8	0.8	22	22	53.88	53.89
65	INISHTRAHULL	8	34	8.0	8.0	22	22	55.44	55.36
66	RATHLIN	14	1487	4.0	4.0	21	21	55.29	55.24
67	JOHN'S COPELAND	10	15	3.8	1.3	22	22	54.67	54.65
68	LAMBAY	15	238	4.5	4.5	25	25	53.48	53.48
69	ALDERNEY	27	794	14.0	14.0	67	67	49.71	49.71
70	HERM	18	130	40.0	7.5	67	24	49.45	49.52
71	SARK	26	516	36.5	20.0	67	36	49.39	49.52
72	GUERNSEY	24	6355	46.5	11.0	67	36	49.41	49.52
73	JERSEY	36	11621	25.0	25.0	67	67	49.16	49.39

S: number of species; **A**: island area (ha); **I1**: isolation of island from nearest mainland source in France, mainland Britain or Ireland (km); **I2**, isolation of island from nearest source with at least an equivalent number of species (km); **FS1** size of faunal source within 50 km of the nearest point on the mainland of France, Britain or Ireland; **FS2**: size of faunal source at nearest source with at least an equivalent number of species; **L1**: latitude of island's most southerly point; **L2**: latitude of nearest point on the mainland.

Table 2. Correlations between the number of butterfly species on 73 British and Irish islands and geographical variables. Upper triangle: Spearman correlations (r_s): lower triangle Pearson correlations (r) for normalized data.

	S	A	I1	I2	FS1	FS2	L1	L2
S		0.226 NS	–0.409***	–0.123 NS	0.668***	0.745***	–0.609***	–0.598***
A	0.212 NS		0.147 NS	0.121 NS	–0.284*	–0.143 NS	0.394***	0.417***
I1	–0.445***	0.187 NS		0.521***	–0.245*	–0.614***	0.238*	0.237*
I2	–0.153 NS	0.179 NS	0.543***		–0.128 NS	–0.089 NS	0.126 NS	0.133 NS
FS1	0.642***	–0.256*	–0.198 NS	–0.079 NS		0.779***	–0.933***	–0.932***
FS2	0.787***	–0.168 NS	–0.607***	–0.116 NS	0.782***		–0.685***	–0.680***
L1	–0.612***	0.353**	0.236*	0.149 NS	–0.931***	–0.703***		0.997***
L2	–0.538***	0.369***	0.206 NS	0.124 NS	–0.928***	–0.672***	0.994***	

***, $P < 0.001$; **, $P < 0.01$; *, $P < 0.05$; NS, not significant.

See Table 1 for variable names.

- isolation from the nearest source of equivalent size (**I2**);
- the number of species within 50 km of the nearest mainland source (**FS1**);
- the number of species at the nearest equivalent source (**FS2**);
- island latitude (**L1**);
- latitude of the mainland source (**L2**);
- isolation from the nearest point on the continent of Europe (**C**, km) is also referred to below, though this variable is excluded from Tables 1 and 2.

The number of species (S) is taken to be the sum of species recorded on each island. We have excluded 27 records for fourteen islands; these include questionable observations and species indicated in the literature to be extinct more than 30 years ago. Island species' numbers correlate significantly with all variables except isolation from the nearest source (**I2**). The most important predictor of species' numbers is the size of the faunal source (**FS2**: $r_s = 0.74$; **FS1**: $r_s = 0.67$; Fig. 3). Correlations with latitude are also substantial (**L1**: $r_s = -0.60$; **L2**: $r_s = -0.59$) as is that with isolation from the Continent (**C**: $r_s = -0.54$). The lowest correlations are for isolation from the nearest mainland source (**I1**: $r_s = -0.39$) and island area (**A**: $r_s = 0.25$), that is with the exception of **I2**.

Sixteen of the twenty eight Pearson correlations (for normalized data; twenty-one of the Spearman correlations) among predictors are significant at $P < 0.05$; thirteen are significant at $P < 0.001$. Particularly high correlations occur between latitude and isolation from the Continent (**L1** with **C**: $r = 0.83$; **L2** with **C**: $r = 0.85$). The number of species at the nearest mainland source (**FS1**) also correlates very highly with latitude (**L1** and **L2**: $r = -0.93$), isolation from the Continent (**C**: $r = -0.88$) and, though somewhat less, with the number of species at the nearest equivalent faunal source (**FS2**: $r = 0.78$). The correlations of the size of the nearest faunal source (**FS2**) with latitude (**L1**, **L2**: $r = -0.67$ to -0.70) and isolation from the Continent (**C**: $r = -0.62$) are more modest but still have substantial amounts of variance in common, as do correlations between the isolation measures (**I1** with **I2**: $r = 0.54$) and between isolation from the mainland and the size of the nearest faunal source (**I1** with **FS2**: $r = -0.61$). Other correlations tend to be small ($r^2 = < 15\%$). Of the twelve correlations that prove not to be significant at $P < 0.05$, six involve isolation from the nearest equivalent source (**I2**), four involve island area (**A**) and three involve isolation from the nearest mainland source (**I1**).

Intercorrelations among predictors (Table 2) largely explain the order and relative size of some of the correlations with species' richness. For example, the substantial correlation of species' richness with isolation from the Continent (**C**) owes almost entirely to the latter's high correlation with latitude and is of little contemporary biological significance. The biological significance of latitude, as a surrogate for climatic parameters, has been discussed fully elsewhere (Dennis, 1977, 1993). It is also noteworthy that British and Irish islands become significantly larger and more isolated further north. Even more important from the vantage of interpreting the correlation and regression parameters is that these larger and more isolated islands further north have access to much smaller numbers of potential colonists. Because of the substantial correlations between

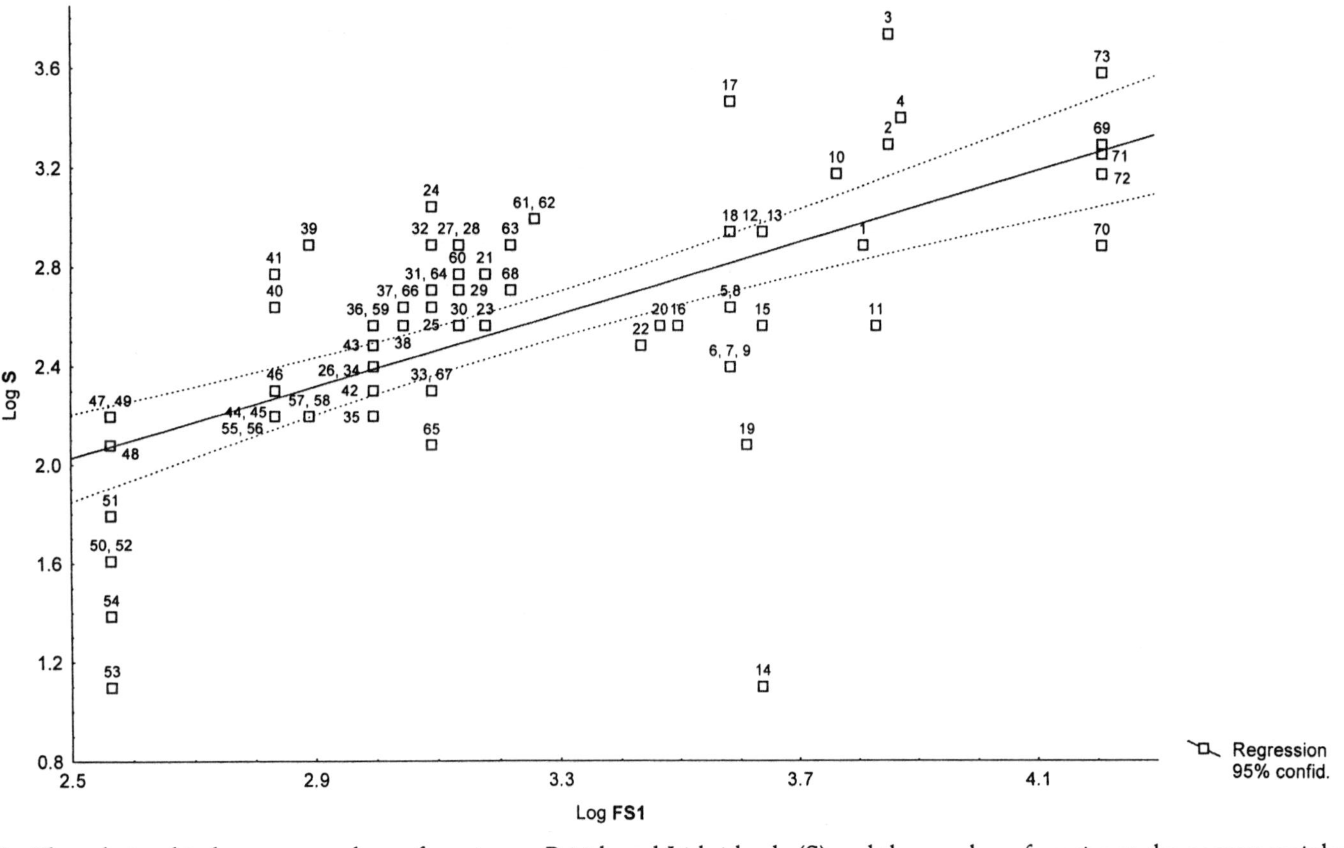

Fig. 3. The relationship between numbers of species on British and Irish islands (S) and the number of species at the nearest mainland source (**FS1**). See Table 1 and Fig. 1 for names and locations of islands.

$$\log \mathbf{S} = 0.199 + 0.73 \log \mathbf{FS1};\ r = 0.64,\ P < 0.001$$

island species' richness and the size of faunal source and latitude, the question of the importance of area and isolation can only be fully addressed by first removing the effects of the size of the faunal source and underlying influences such as latitude and climate. To do this we have applied what in effect amounts to a series of step-wise multiple regression models, in which the entry of variables is predetermined. The format we have selected for analysis enables the influence of a dominant variable(s) to be removed and determination of the extent to which variables targeted for comparison account for the residual variation *independently* of one another. Using this particular step-down procedure, the maximum possible contribution of island area and island isolation to species' numbers can be directly compared with one another.

A summary of the parameters from the regression analyses is given in Table 3. Of the island area and isolation measures (**A**, **I1**, **I2**), only isolation from the nearest mainland point (**I1**) is significant ($r^2 = 0.2$), whereas both faunal sources (**FS1**, **FS2**) and latitude (**L1**) are confirmed as accounting for a substantial portion of the variance. However, the residuals from the regressions of species' richness on these three latter variables (**RES1** for **FS1**, **RES2** for **FS2**, **RES3** for **L1**) are significantly correlated with island area ($r^2 = 0.24$–0.31) and isolation from the mainland (**I1**: $r^2 = 0.15$–0.17), but not with isolation from the nearest equivalent source (**I2**: $r^2 \approx 0$). Much the same picture emerges for residuals from the regression of island species' richness with island latitude and the size of faunal source (**RES4** for **L1** and **FS1**; **RES5** for **L1** and **FS2**). Area again accounts for more residual variance (**RES4** and **RES5** on **A**: $r^2 = 0.26$–0.37) than isolation (**I1** on **RES4**: $r^2 = 0.17$; **I2** on **RES5**: $r^2 \approx 0$), a result that belies the difference in order of simple correlations between species' richness, island area and isolation.

The significance of the size of faunal source (both **FS1** and **FS2**) and latitude, in accounting for species' richness on islands, is to be expected (Dennis, 1992). Mainland points account for more than 41% of the variance and closer sources (larger islands), which may act as reservoirs and stepping stones, account for more variance still (62%). The size of the faunal source at points on the mainland (**FS1**) is intimately linked to latitudinal gradients in species' richness; thus the high correlations between faunal source with latitude (**FS1** with **L2**: $r = -0.93$; **FS2** with **L1**: $r = -0.69$). The relationship between the latitudinal gradient in butterfly species' richness and climatic gradients, especially summer sunshine and temperatures, is well known and largely understood (Turner, 1986; Turner, Gatehouse & Corey, 1987; Dennis, 1977, 1992, 1993; Dennis & Williams, 1986). Simply put, the lower summer sunshine levels and temperatures of northern Britain sustain fewer butterfly species than do the higher sunshine and temperature levels of southern Britain. Thus, islands at higher latitudes have smaller banks of species as potential immigrants and colonists. Compared to the size of faunal sources, and the influences underlying faunal source, island area and isolation account for very little of the variance in island species' numbers. Area and isolation (**I1**) may account for as little as 26% and 17% respectively of the *residual* variance. Maximally, area is unlikely to account for more than 14% of the total variance (37% of the residual 37.3% of the variance from the regression of **S** on **FS2** and **L1**) and isolation (**I1**) as little as 20% of the total variance (from the

Table 3. Summary of regression analyses of numbers of butterfly species on 73 British and Irish islands with selected geographical variables.

a) regressions of the number of species on 73 islands and of key residuals on the geographical variables.

Dependent variable	Independent variable	Constant	Slope	r^2	$F_{(1,71)}$	P	Residual
log S	log **A**	–	–	–	3.35	NS	
	log **I1**	2.96	–0.157	0.20	17.57	<0.001	
	log **I2**	–	–	–	1.70	NS	
	log **FS1**	0.20	0.731	0.41	49.71	<0.001	**RES1**
	log **FS2**	0.42	0.716	0.62	115.87	<0.001	**RES2**
	log **L1**	24.26	–5.429	0.38	42.51	<0.001	**RES3**
RES1	log **A**	–0.50	0.070	0.24	22.54	<0.001	
RES1	log **I1**	0.28	–0.112	0.17	14.77	<0.001	
RES2	log **A**	–0.46	0.064	0.31	32.27	<0.001	
RES2	log **I2**	–	–	–	0.71	NS	
RES3	log **A**	–0.57	0.080	0.29	29.41	<0.001	
RES3	log **I1**	0.05	–0.031	0.14	12.01	<0.001	
RES3	log **I2**	–	–	–	0.44	NS	

b) multiple regressions of the number of species on 73 islands on the geographical variables and between residuals and the geographical variables.

Dependent variable	Independent variable	Constant	Slope	R^2	$F_{(2,70)}$	P	Residual
log S		4.51					
	log **FS1**		0.613				
	log **L1**		–0.984	0.41	24.67	<0.001	**RES4**
log S		4.72					
	log **FS2**		0.642				
	Log **L1**		–1.021	0.63	58.77	<0.001	**RES5**
				r^2	$F_{(1,71)}$		
RES4	log **A**	–0.52	0.072	0.26	24.72	<0.001	
RES4	log **I1**	0.27	–0.110	0.17	14.18	<0.001	
RES5	log **A**	0.50	0.070	0.37	41.67	<0.001	
RES5	log **I2**	–	–	–	0.55	NS	

In (a), island species' richness (**S**) is first regressed against single predictors. Three residual variables, **RES1** to **RES3** (from **S** regressed with **FS1**, **FS2** and **L1** respectively), are then regressed against island area (**A**) and isolation (**I1**, **I2**). In (b), island species' richness (**S**) is first regressed against the joint predictors, size of the faunal source and island latitude (**FS1**, **L1**). The two residual variables **RES4** and **RES5** are then regressed against island area (**A**) and isolation (**I1**, **I2**).

For variable codings see Table 1.

simple correlation of **S** on **I1**). The more likely figure for isolation (**I1**) is 9.8% of the total variance (16.7% of the residual 58.7% of the variance from the regression of **S** on **FS1** and **L1**).

The figures for island area and isolation may seem surprising. The impression gained from the regression parameters is that the influences of island area and island isolation on butterfly migration, colonization and/or survival are rather weak. This may have a great deal to do with the size and isolation of islands included in this survey. The mean size of islands is 16,221 ± SE 4,176 ha; only twelve islands are less than 1 km^2 and only three are less than 10 ha. Thus, the area of islands compares very favourably with the minimum population area on which species are known to persist for several years (Warren, 1992). Although the mean isolation of islands from the nearest mainland point is 28.7 ± SE 5.0 km, the mean distance to the nearest equivalent source is 7.0 ± SE 1.1 km; 66% of islands are less than 5 km away from a source with, at least, an equivalent number of species. This sea distance is well within the migration capacity of many resident British species (see section I.7). Nevertheless, it is important to consider that these results do not provide an unbiased estimate. First, data on the islands are cumulative numbers of species recorded over a long period (>50 years); they may not reflect the number of species that would be obtained by an equal effort survey over a single season. Secondly, there is an obvious bias to having data on larger islands; the greater number of observations made on larger islands are more likely to have scored the presence of the four indicator species used for determining island selection. As more smaller islands are surveyed, regression parameters for area and isolation could increase in size and significance.

The present results can be compared to previous analyses by Dennis (1977), Hockin (1981) and Reed (1985). An assessment of these earlier surveys was made by Dennis (1992), in which data corrected from Reed (1985) were re-analysed (Table 4). Analyses of island records has been applied to increasing numbers of islands; Hockin included 29 islands in his calculations, whereas Reed used data from 52 islands. However, both data sets proved to be grossly inaccurate, as

Table 4. Correlations (Pearson r) from previous analyses between the numbers of species of butterflies on British and Irish islands and five significant predictor variables.

Predictor variable	Hockin	Reed	Corrected data from Reed
Island area	0.22	0.36	0.19
Isolation	–0.67	–0.42	–0.55
Number of species at nearest source	0.77	0.46*	0.65
Number of plant species	0.72	0.73	0.80
Latitude	–0.74	–0.39	–0.69
Number of islands	29	52	45

The British mainland and Ireland are removed from all analyses by Hockin (1981) and Reed (1985). In both analyses substantial errors in species' numbers occur. All variables have been transformed to $\log_{10}$ by Hockin, but only area and distance by Reed.

* Calculated from available data. When the effect of latitude is removed, area accounts for more of the residual variation (0.49) than does isolation (–0.23).

became evident in a re-analysis of a selection of 45 islands from Reed's list (Dennis, 1992). The correlations produced results more in line with Hockin's findings. It has to be said that some of the disparity in the results of Hockin's and Reed's surveys can be accounted for in the number of islands entered into analyses and the numbers of species scored for them. In Hockin's survey, the small sample of larger islands tends to be biased to Scotland; this difference is not significant between Hockin's and Reed's data ($\chi^2_{(1)} = 2.42$, $P > 0.1$). However, compared to the current file of islands there is heterogeneity (Hockin's data: $\chi^2_{(1)} = 5.0$, $P < 0.05$; Reed's data: $\chi^2_{(1)} = 0.8$, $P > 0.3$). In both surveys, there are substantial deficits in the number of species based on the data available at the time of each survey (e.g., Jersey in Reed; Coll in Hockin). In both surveys, too, some islands were included that to this date have been inadequately surveyed (e.g., Clare, Dursey and Aran in Reed). Some disparity in the results may also derive from the different treatment of variables (i.e., variable transformations; combinations of variables entered into multiple regression models). Moreover, slightly different variable suites have been employed. Both Hockin and Reed investigated the influence of island area, isolation from the nearest mainland source, island elevation, latitude, number of soil types and number of plant species. Hockin additionally included the size of the faunal source (viz., number of species breeding within 25 km of the nearest point on the mainland), whereas Reed added longitude.

The disparity between these earlier surveys and the present one primarily focuses attention on the relative contribution of island area, island isolation and the size of faunal sources to species richness on islands. In both Hockin's and Reed's surveys, as well as the re-analysis of Reed's data, simple correlations would indicate that area *per se* holds less significance for species' richness than isolation. From multiple regression equations, Hockin finds that species' richness is mainly explained by the combination of the size of the faunal source and isolation ($R^2 = 0.69$). On the other hand, Reed maintained that number of plant species, a surrogate for habitat diversity, is the main determinant of butterfly species' richness ($r^2 = 0.53$), the residual variance accounted for by latitude ($r^2 = 0.14$) and isolation ($r^2 = 0.06$). The re-analysis of Reed's data demonstrated that when the effect of latitude on size of the faunal source was removed, island area accounted for more of the residual variation ($r = 0.49$) than does isolation ($r = -0.23$).

The results of the present survey confirm some of the findings of Hockin (1981) and Reed (1985), but contest others. For instance, they confirm the importance of the size of the faunal source (see Dennis, 1977), which Reed excluded from consideration. They also confirm the rather weak correlations for area and isolation; this corresponds to findings by Reed, but contests the high correlation for isolation in Hockin's survey. However, Hockin's findings may have been adversely affected by the small sample of islands. Reed attributed species' richness to habitat diversity, which he purported to be better measured by the number of plant species than by island area. The interpretation of this finding has already been challenged (Dennis, 1992). Although the number of plant species may in some degree measure island habitat diversity, it is subject to the same island biogeography influences (viz., migration, colonization and extinction) as the butterfly faunas (e.g., plants on area and isolation: $R^2 = 0.38$). Secondly, the number of plant species almost certainly also reflects a response to environmental gradients, such as sunshine and temperatures. Thirdly, there is some indication that both plants and butterflies in Reed's survey may have been inadvertently influenced by similar sampling deficiencies; islands poorly surveyed for butterflies may tend to be those inadequately surveyed for plant species, particularly as surveys prior to 1960 often collected data on a wide range of organisms. Finally, the number of plant species does not measure the suitability of habitats for butterflies directly, as the geographical ranges of hostplants *per se* do not limit butterfly species throughout Britain (Dennis & Shreeve, 1991). They may, however, measure environmental suitability indirectly which can influence butterfly biogeography.

As a final point on the present results, we would draw attention to the fact that they are based largely on unsatisfactory data and that it has been necessary to assess records that have been accumulated over a period of years. Ideally, analyses would have access to data from shorter-term synoptic surveys in which the status of species on islands is clear. However, basic patterns and trends identified in the current survey are distinctive and highly significant and are unlikely to be deficient in any future survey.

4. RELATIONSHIPS AMONG ISLANDS FOR BUTTERFLY FAUNAS

An analysis of factors underlying species' richness does not take account of the assemblages of species on islands and of the taxonomic relationships between island faunas. Techniques for doing this include multivariate statistical routines of ordination and clustering (see section I.2). For this purpose, it is valuable to generate comparative models for the distribution of units, in this case for islands.

A. The influence of geography

The most obvious model for inter-island comparisons is that of geographical position, with the expectation that islands in close proximity will have butterfly faunas more similar to one another than islands further apart. Ordination and clustering plots of the faunal similarities between islands (island matrix: S_J) can be compared directly with the geographical location of islands visually or by regressing similarities between islands on their faunal content against the geographical distances between them. A graphical comparison was carried out in an earlier work for a limited number of islands (Dennis, 1977: 78, 196). It was shown that a non-metric ordination plot for the islands reproduced some geographical order, though all islands were 'driven' northwards and smaller islands forced to the periphery of the plot. A correspondence of an ordination of islands on their faunal content with geography would imply that systematic influences are at work on the distribution of species. Obvious candidates are:

(i) isolation which affects migration of species between units;
(ii) gradients in resources and conditions that affect the adaptability of species.

With regard to isolation the presence of a proximate faunal source for an island, and any variation in the size of the islands, could significantly disrupt any geographical order. The exceptions to this are if isolation from a faunal source and variation in the size of islands correlate with the geographical ordering of the islands.

In the case of the British islands there is some interaction among island area, isolation and the size of the faunal source; islands become larger and more isolated further north and the faunal source decreases in size. In the absence of faunal gradients across an archipelago, increasing size and increasing isolation would be thought to go some way to cancelling each other out in their effect on species' richness. However, the impact on faunal content is not so easily predicted; inter-island relationships would depend on distinctions for migration, colonization and survival among species. Data on ecological parameters for resident species (see Table 8) in Britain would suggest that such distinctions exist and are worth exploring. Yet, whether the increments in island size and isolation northwards are of an order sufficient to exploit ecological and behavioural differences among butterfly species is another matter.

In Britain, the factors most likely to generate geographical order in inter-island comparisons are environmental gradients. Britain has very distinctive climatic, geological and topographic gradients which interact to have a dramatic impact on species' ranges and species' richness (Dennis, 1977, 1993; Turner *et al.*, 1987). The gradient in conditions effectively filters out potential colonists to the north-west of Britain and may thus significantly influence the pattern of inter-island affinities based on butterfly species, generating nested species-subsets from north to south. In turn, islands will tend to have subsets of species from the nearest faunal sources, but influenced by the stochastics of colonization and extinction. The smaller the island fauna, the more probable it is that random effects will influence species' composition. Thus, islands, especially small isolated ones, will tend to draw away from mainland regions based on species' associations. In the current work, we test the influence of geographical location on the ordination of islands in two ways. First, we determine the degree to which geographical location provides a suitable model from which to ordinate islands on species' associations. Second, we compare affinities among islands, based on their butterfly faunas, with their geography by regression analysis.

For the first task, we apply two initial configurations for islands entered into non-metric multidimensional scaling routines:

(i) the default Guttman-Lingoes initial configuration;
(ii) geographical positions, the longitude and latitude of islands (Fig. 4).

Some detail can be found on non-metric scaling in section I.2, including a brief discussion of initial configurations. What is important to consider in the following comparison is that an initial configuration in two dimensions based on the geography of the islands has a Kruskal stress = 0; that is, there is no difference between the distances among units in the initial plot and the distances taken from an atlas. The two resulting plots are virtually identical for most units. However, differences in the placement of islands do exist. For example, St Johns Copeland switches allegiance in the plots from being near Scalpay (Guttman-Lingoes initial configuration) to being near Tresco (geographical initial configuration). The correct placement (nearer Tresco) is determined by extending the ordination to six dimensions. Even in three dimensions, it is evident that the Irish islands separate from those in the Isles of Scilly group. The use of geography also reduces the Kruskal stress value by 1% from 0.11 to 0.10 (Fig. 4). The current ordination for 73 islands reproduces the basic features of the earlier plot from 1977 for 19 islands (Dennis, 1977). First, there is a semblance of geographical order along dimension 1. Southern regions and islands lie to the left of the plot and northern islands, such the Shetland group, are to the right. Second, all islands have been shunted 'northwards' into positive values on dimension 1 compared to the mainland zones used as a datum level, the continental group of France, Belgium and Holland, and the four latitudinal belts of the British mainland. In fact, the two most southerly British mainland zones have been similarly displaced compared to

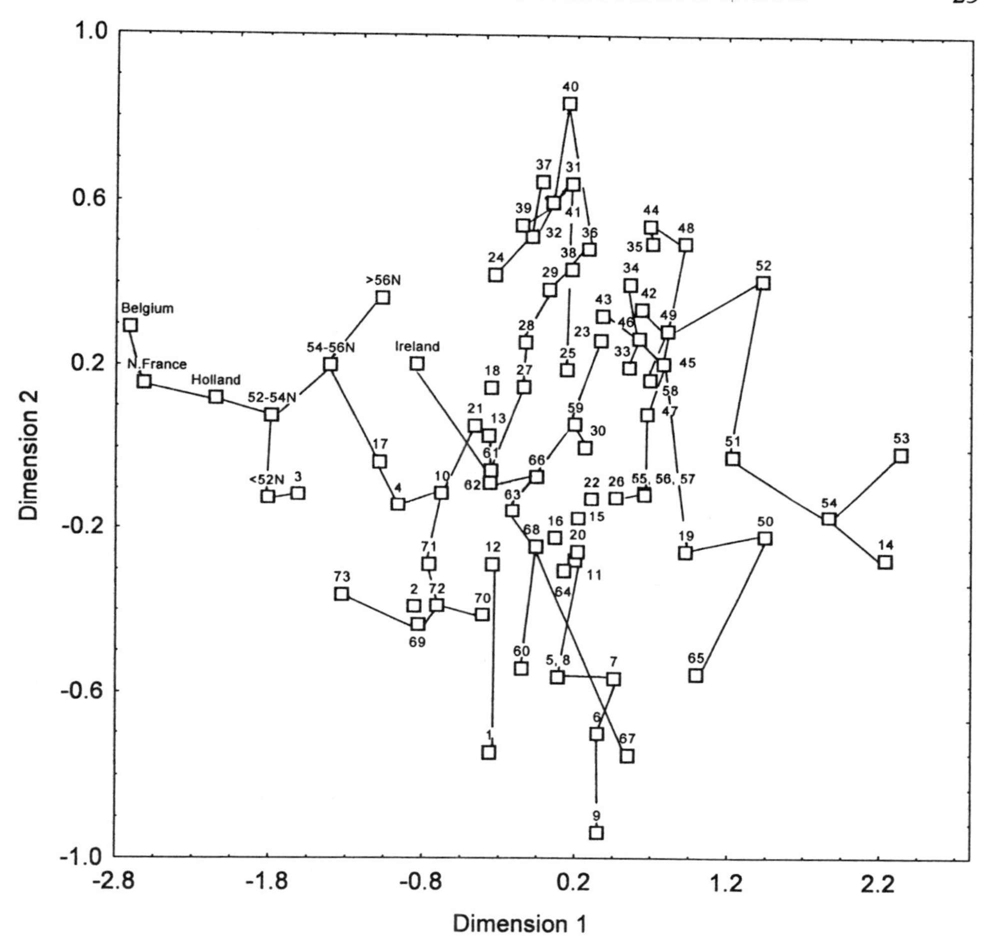

Fig. 4 Non-metric scaling plots (2 dimensions) for British and Irish islands based on their similarities (S_J) for butterfly species. Island geography (longitude and latitude) is used for the initial configuration. The islands are linked by a minimum spanning tree from single linkage clustering (Guttman's K = 0.108; Kruskal phi = 0.102). The position of Cramond and Inchcolm is the same as that as May, and St Mary's the same as that of St Martin's. See Table 1 for names of islands.

the continental regions though northern Britain has been drawn south of the islands and closer to the Continent. Third, maximum displacement has affected those islands with the smallest faunas (e.g., Grassholm; Puffin Island).

Unlike the earlier plot published in 1977, there is a much greater spread of islands on the second dimension. Islands at opposite extremes on dimension 2 have similar species' richness but clearly differ in affinity. There is a tendency for islands having positive values on dimension 2 to be northern islands, particularly from the Western Isles of Scotland, and for those having negative values on dimension 2 to be from southern Britain. The latter islands potentially have greater access to species which do not generally range as far north as Scotland. Exceptions do exist, notably Inishtrahull and Johns, Copeland, islands off northern Ireland, which have closer affinity to islands from the Isles of Scilly group. It is worth pointing out that some butterfly species (e.g., *Argynnis paphia* and *Leptidea sinapis*) are more abundant in northern Ireland than they are at equivalent latitudes on the British mainland. As a word of caution, too much reliance cannot be placed on the order of islands on the second dimension in the two dimensional solution (Fig. 4) as the main contribution to Kruskal stress occurs in the relative position of some of the Irish islands on this dimension.

In the second test of the influence of geography on island butterfly faunas, we regress affinities among islands for their butterfly faunas on their mean isolation from one another. Affinities are mean Jaccard coefficients ($\bar{S}_J$), that is, the mean similarity of any one unit with every other unit but itself. Average isolation is the mean geographical distance on an island to all other islands. The regression analysis reveals that geography may significantly influence affinity among islands ($r = -0.54$, $P < 0.001$; Table 5; Fig. 5), affinity declining with mean isolation, as would be expected. However, the large residuals for some islands (e.g., Grassholm; Isle of Wight) indicate that other important factors also influence affinities among islands.

Cluster analysis largely confirms the validity of the ordination produced by non-metric scaling. The links from nearest neighbour clustering are illustrated on the non-metric plot (Fig. 4). They produce a typical branching pattern in which, for the most part, nearest neighbours in the plot have been fused, faithfully reproducing their proximity in the similarity matrix and in the ordination.

Table 5. Correlations (Pearson r) of geographic distances, affinities and modelled affinities for 73 British and Irish islands (see text for explanation).

	MEANDIST	$\bar{S}_J$	$\bar{S}_{JRAN}$	$\bar{S}_{JREG}$
$\bar{S}_J$	–0.542***			
$\bar{S}_{JRAN}$	–0.057 NS	0.173 NS		
$\bar{S}_{REG}$	–0.362**	0.453***	0.673***	
$\bar{S}_{JLOC}$	–0.363**	0.502***	0.685***	0.996***

***, $P < 0.001$; **, $P < 0.01$; *, $P < 0.05$; NS, not significant.

MEANDIST Mean geographic distance between islands.

$\bar{S}_J$ Mean similarity between islands, based on actual species present.

$\bar{S}_{JRAN}$ Mean similarity between islands derived from 25 simulations where species for each island are selected at random from the mainland species pools of sources in Ireland, Britain and northern France.

$\bar{S}_{JREG}$ Mean similarity between islands derived from 25 simulations where species for each island are drawn at random from one of six sources in Ireland, northern France and Britain (<52°N, 52–54°N, 54–56°N, >56°N) according to the location of each island.

$\bar{S}_{JLOC}$ Mean similarity between islands derived from 25 simulations where species are drawn at random from the species pool at the nearest faunal sources (50 km squares) on the mainlands of Britain, Ireland and northern France.

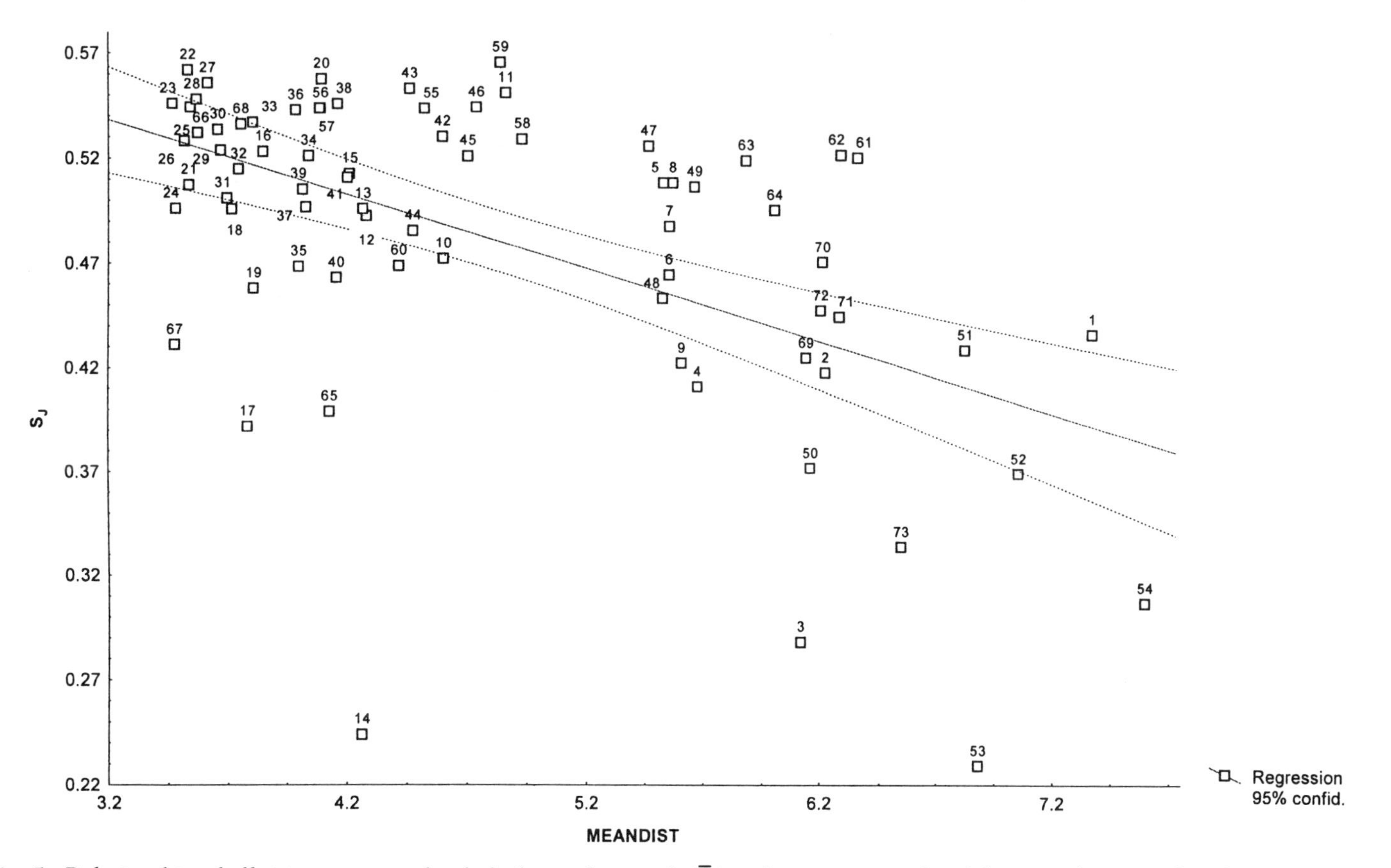

Fig. 5. Relationship of affinities amongst islands for butterfly records ($\bar{S}_J$) with mean geographical distances between the islands. See Table 1 for names of islands.

$$\bar{S}_J = 0.65 - 0.35\text{MEANDIST};\ r = -0.54,\ P < 0.001$$

Exceptions do exist, usually extreme units on the margin of the plot. For example, Sheppey links up with Skomer towards the centre of the 'map'; similarly, John's Copeland links up with Lambay, as does Puffin with Inner Farne, rather than with nearby islands in the plot. However, the differences in the size of similarity coefficients is small, and the placement of such islands in the plot is clearly indicative of their multiplicity of affinities with other units.

Complete and UPGMA clustering produce somewhat similar dendrograms and largely agree for the classification of low and mid-order clusters. Very few tight clusters exist and these are not highly differentiated. Only islands in two small groups have identical collections of species:

(i) May, Cramond and Inchcolm in the Firth of Forth;
(ii) St Martin's and St Mary's in the Isles of Scilly.

Both dendrograms have a tendency to chaining, with pairs or triplets of closely related islands linking up loosely to single islands or to other similarly sized clusters. The impression gained from this is that the ordination plot (Fig. 4) provides a more useful base from which to determine relationships than the dendrograms. Consequently, the latter are not illustrated. Nevertheless, some generalizations can be made:

(i) There is some geographical order in lower ranked clusters; for example, the seven southern mainland zones on the Continent and in Britain are clumped, as are islands from the Shetland group, and many of the Hebridean islands.
(ii) High-ranked clusters fusing at low similarity levels do not make up homogeneous groups of islands based on geography; there is much intersection of clusters on the geography of elements. For example, Jersey and the remaining Channel Islands are found in different clusters. The Irish islands are also dispersed, separating the larger from the smaller ones.
(iii) Small islands with small faunas tend to have identical affinities in the complete linkage and the UPGMA clustering dendrograms (e.g., Grassholm with Shetland and Puffin Island with John's Copeland), but the two solutions are far from identical.

Thus, differences exist for low-order as well as high-order clusters. For example, Ireland is placed with the larger Irish islands and the Isle of Man in the complete linkage dendrogram, but with the northern regions of mainland Britain and Anglesey in the UPGMA dendrogram. Overall, although there are a good number of similarities between the solutions from the cluster analyses, there are also obvious differences; neither has revealed prominent reproducible homostats (tight clusters) of islands.

B. The influence of species' richness and faunal sources

Another model that can be compared to the species' assemblage data for islands accounts for differences in species' richness. The relationships among islands will be influenced, to some extent, by the size of the island faunas. The contribution of faunal size to the similarities among islands can be assessed by randomly allocating a quota of species to islands from the pool of species in the whole region, the quota limited by the species' richness of each island. The affinities (mean Jaccard coefficients, $\bar{S}_J$) among islands, based on data for their butterfly faunas, can then be compared to affinities between them calculated for their random array of species. The correlation between them ($r = 0.173$; $P = 0.143$; Table 5) indicates that faunal size *per se* does not have a significant effect on affinities among islands. Higher affinities would be expected for larger islands. The sign is in the right direction, but even if it were significant would account for less than 3% of the variance in island affinities.

The above model, testing for differences in species' richness, can be modified to take account of the size of the species' pool accessible to different islands. This would effectively measure the influence of a regional faunal source. Two versions of faunal source have been used in this simulation to compare the potential relative influence of local versus regional faunal sources of adjacent mainlands on island faunas. In the first case, species are allocated to islands from one of six regional species' pools: northern France, the British mainland (four regions divided at 52°, 54° and 56° N) and Ireland. In the second case, species are allocated to islands from species' pools within 50 km^2 zones at the nearest mainland source for the Continent, the British mainland and Ireland. Again, the quota for each island is limited by the number of species recorded for it. In both models affinities are derived from 25 simulations. Observed affinities among island faunas can then be regressed against equivalent affinities for the two models. In each case the correlation is highly significant, though interestingly enough, slightly lower than the correlation between observed affinities and mean isolation (Table 5). What is particularly revealing is that randomly allocating species from local mainland sources only marginally improves the estimate compared to that of randomly allocating them from regional sources (i.e., regional sources, $r = 0.45$, $P < 0.001$; local sources, $r = 0.5$, $P < 0.001$). The inference is that local faunal sources largely mirror regional species' pools and this is confirmed by the correlation between the two sets of affinities for islands modelled from local and regional sources ($r = 0.996$, $P < 0.001$).

In detail, there are significant discrepancies among islands in the regression of observed affinities for islands against those modelled on mainland regional and local sources. Those that have a higher observed affinity, than would be expected from a regression against a random allocation of local mainland species, tend to be islands with low species diversity, many of them northern and western islands small in area. This suggests that they have a more homogeneous array of species than would be expected from a stochastic model and are colonized by a rather limited but characteristic assemblage of species than are islands in southern Britain. The exceptions to this are islands with very low species' richness such as Grassholm, Foula and Yell, which have a lower affinity than might be expected from a random allocation of species. Also, with distinctly lower affinities than expected from the model are islands with the highest species' richness (e.g., Isle of Wight, Jersey, Anglesey). These islands are mainly in southern Britain and are more heterogeneous in faunal composition compared to other islands.

In conjunction with data on species' richness, the inference is that butterfly species' distributions in Britain are nested; faunas on small islands are typically nested subsets of those on adjacent larger islands, and faunas in northern Britain are largely nested subsets of those in southern Britain. This pattern is dictated primarily by environmental factors which are shown to underly the latitudinal gradient in species' richness (Turner *et al.*, 1987; Dennis, 1992, 1993). The degree to which island and regional species' assemblages are nested is considered in the following section.

5. BUTTERFLY ASSOCIATIONS ON ISLANDS

Another way to investigate the influence of species' assemblages on relationships between islands is to assess affinities among species for their presence and absence on different islands. Ordination of the similarities among species (species matrix: S_J) by non-metric scaling produces a plot in two dimensions with low Kruskal stress (Fig. 6; d-hat stress, 0.0917). That this is a good representation of relationships is indicated by the minimum spanning tree from single linkage clustering (Fig. 6a); few distortions are evident in the links between species, although some species do not join with their nearest neighbours in the plots (e.g., *Thymelicus acteon*, *Maculinea arion*, *Aphantopus hyperantus*). The non-metric plot

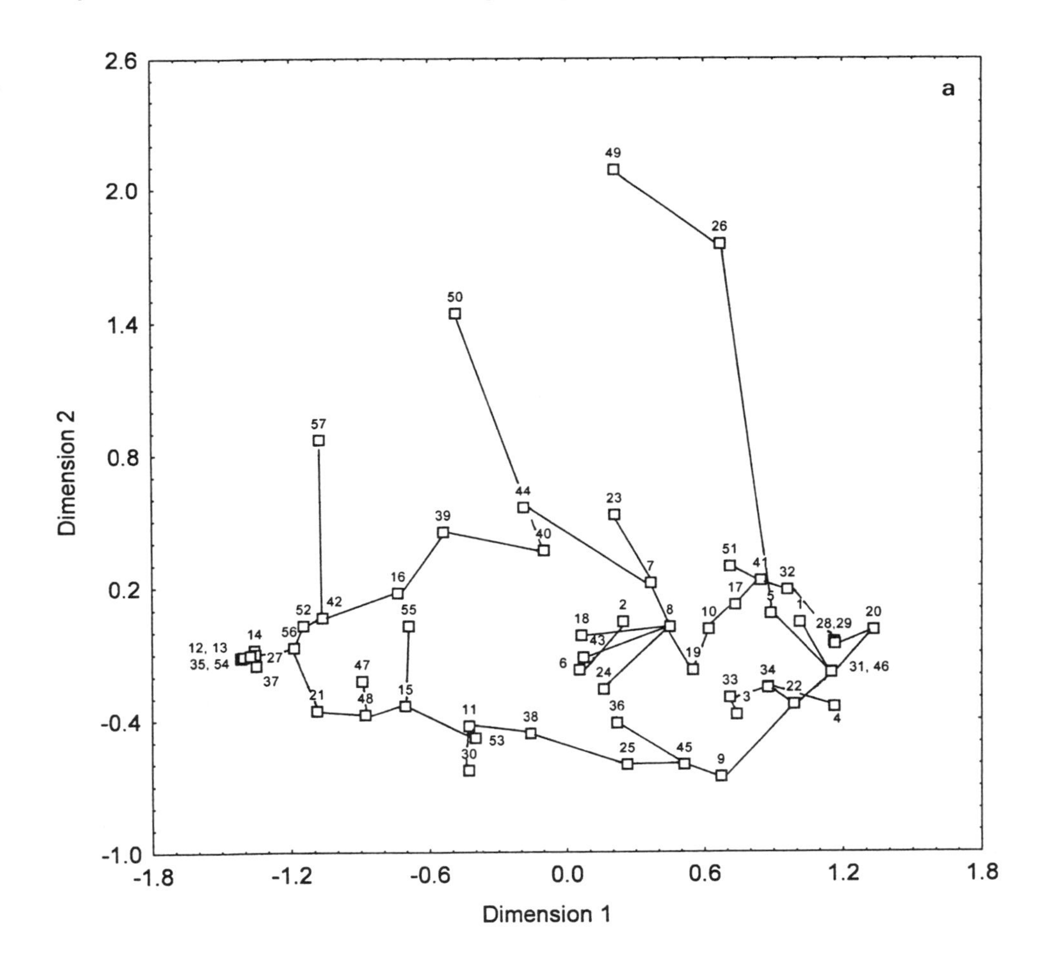

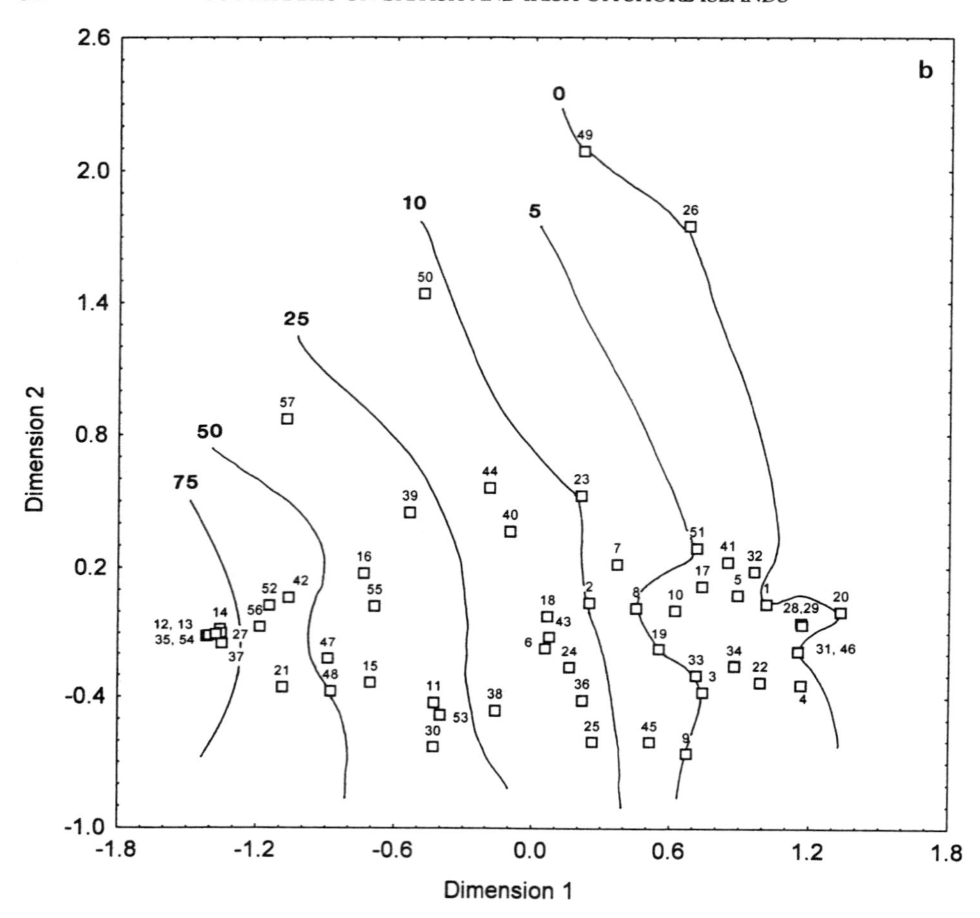

Fig. 6. Non-metric scaling plot (2 dimensions) for butterfly species based on similarities (S_J) for their incidence on 73 islands and eight mainland regions (France, Holland, Belgium, four regions of Britain and Ireland); (a) species are linked by a minimum spanning tree from single linkage cluster analysis; (b) isolines illustrate % incidence on islands. See Table 7 for names of species. (Guttman's K = 0.099; Kruskal phi = 0.092.)

also includes isolines for island incidence (Fig. 6b). In the plot, dimension 1 describes range size in species, such that those with positive values (e.g., *Satyrium pruni*, *Lysandra coridon*) have small ranges, whereas those with high negative values (e.g., *Pieris napi*, *Aglais urticae*) have widespread ranges. Dimension 2 effectively separates northern species with southern range margins within the British Isles (e.g., *Erebia epiphron*, *Aricia artaxerxes* with positive values) from southern species with northern range boundaries (e.g., *Melitaea cinxia*, *Thymelicus acteon* with negative values). The effect is that dimension 1 distinguishes species on the basis of their incidence for islands, whereas dimension 2 separates species found mainly on northern islands from those found predominantly on southern ones.

Bearing in mind the spread of islands on dimension 2 in the non-metric plot of Fig. 4, it would prove of great interest if clusters were to emerge on the plot of affinities among species (Fig. 6). The application of complete clustering and UPGMA clustering does reveal some tendency to produce similar clusters of species (Fig. 7). For example, *Erebia epiphron* and *Aricia artaxerxes* are clearly distinguished from other species; these do not generally occur outside the larger land masses of the Continent, Britain and Ireland, though A. *artaxerxes* has been recorded on Ailsa Craig (Gibson, 1952). There is also some indication of a similar

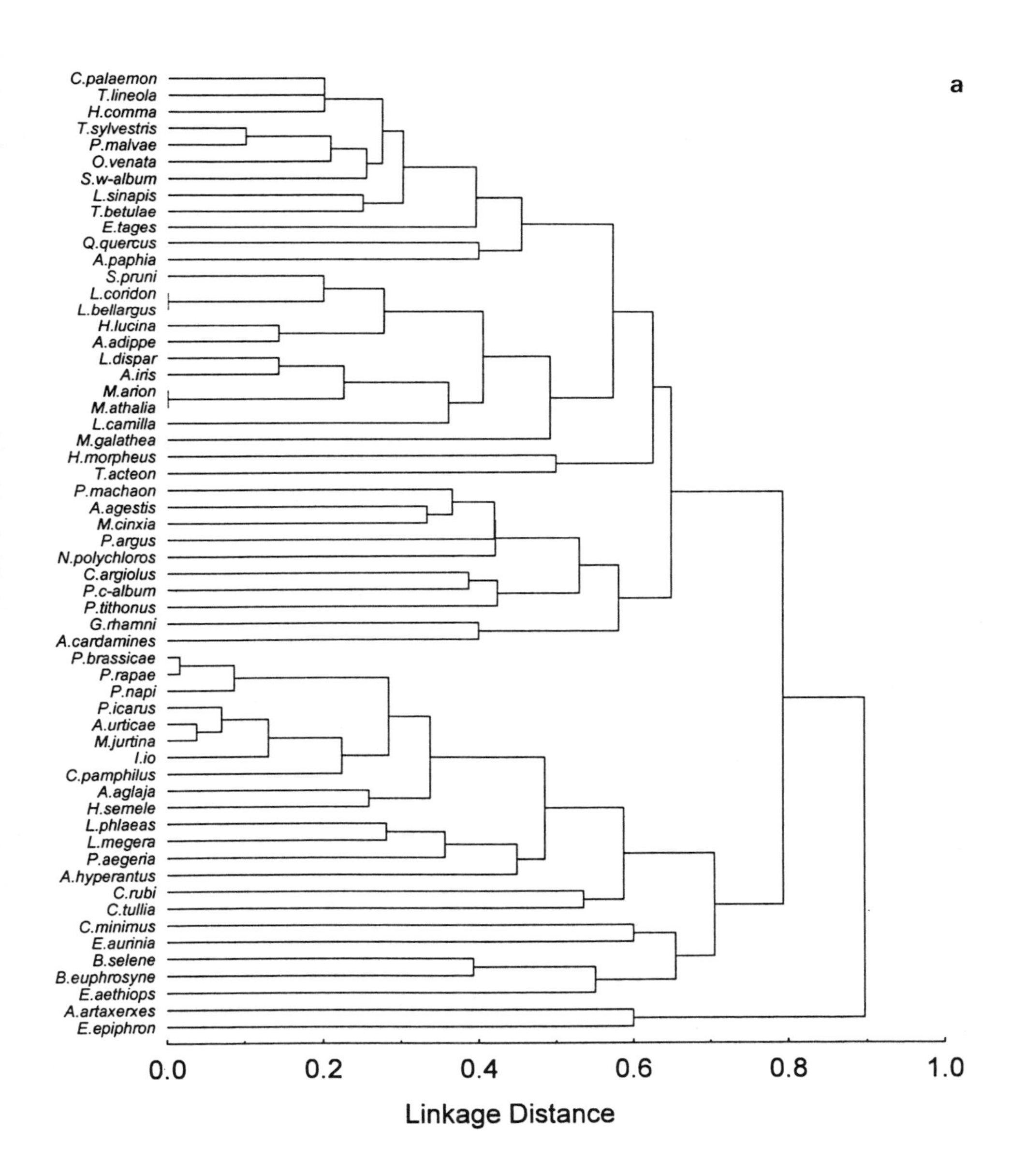

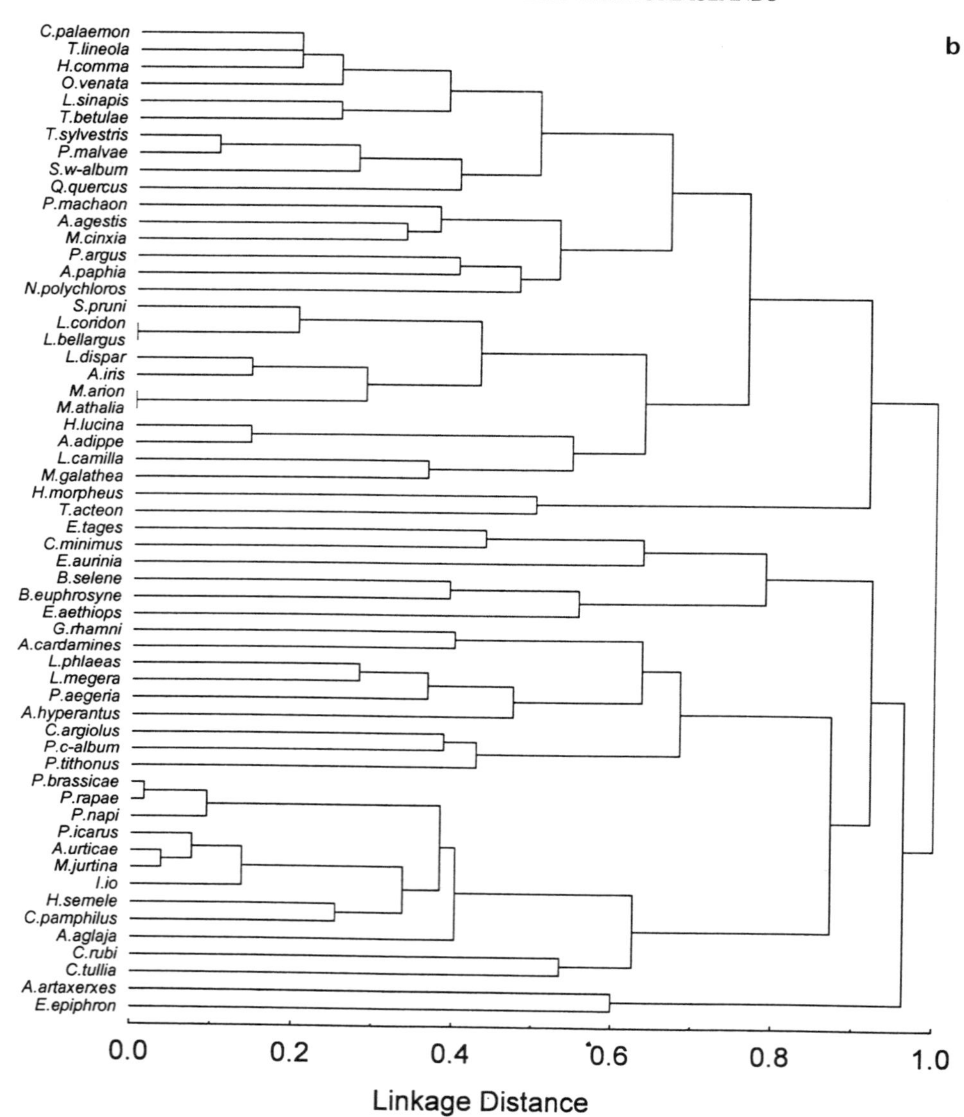

Fig. 7. Cluster analyses for butterfly species based on similarities ($1-S_J$) for their incidence on 73 islands and eight mainland regions (France, Holland, Belgium, four regions of Britain and Ireland): (a) maximum linkage; (b) unweighted pair-group average linkage.

division of the species into two groups, those found frequently on islands and those found infrequently and then on the smaller number of southern islands. All clustering techniques also distinguish a compact cluster of at least 10 species (i.e., *Pieris brassicae*, *P. rapae*, *P. napi*, *Polyommatus icarus*, *Aglais urticae*, *Inachis io*, *Argynnis aglaja*, *Hipparchia semele*, *Maniola jurtina*, *Coenonympha pamphilus*) which have a high probability of occurring on islands throughout the British Isles. Apart from this, there is little evidence for natural clusters in the dendrograms and ordination techniques describe the pattern of relationships better than clustering techniques.

The present results largely confirm the existence of nested-species subsets across islands. In Britain, the number of species on islands increases from north to south, and southern islands tend to accrue, systematically, those species on northern islands. Nevertheless, exceptions do exist. As noted above, in Fig. 6 dimension 2 distinguishes species with southern range limits from those which have northern range limits. In two cases (i.e., *Erebia epiphron* and *Aricia artaxerxes*) of species now found only on the mainland masses, this does not affect the faunal relationships among islands. But, affinities among islands are clearly affected by at least two species with southern range limits that do not occur on southern islands (i.e., *Coenonympha tullia* and *Erebia aethiops*). These two species are clearly not nested within the expanding species' sets of southern islands.

6. PREDICTING BUTTERFLY RECORDS FOR ISLANDS

Data on the islands allow the prediction of:

(i) the numbers of species occurring on islands;
(ii) the likelihood of individual species occurring on particular islands.

The quality of predictions depends largely on the amount of 'explained' variance in the regression models and this, in turn, depends on which variables are used as predictors. In this section, we restrict ourselves to providing an estimate of the numbers of species on islands other than the 73 islands entered into analyses. This may provide some encouragement to those visiting islands to make observations on butterfly species occurring on them (see Appendix 2). Our intention has been to make the prediction of species as simple as possible and one based on readily measured variables. The predictions here are based on the size of the faunal source (**FS1**), isolation of the island (**I1**) and island area (**A**). Island latitude does not significantly add to the prediction. The size of the faunal source on the adjacent mainland and isolation from this faunal source have been chosen in preference to that of the nearest island with an equivalent or larger number of species. Although this does not give quite as high an estimate of 'explained' variance (70% versus 78%), the data for the mainland faunal sources are readily available, whereas it is not for adjacent island sources and it may not be clear which of a number of nearby islands for small target islands is to be used. In using the variables **FS1**, **I1** and **A**, it should be possible for readers to enter data into the equation (see below) for islands not included in Table 6. For completeness, coefficients for the two stepwise multiple regression analyses are given at the end of Table 6; for **S** on **FS1**, **A**, **I1** and **L2**, and **S** on **FS2**, **A**, **I2** and **L1**. No attempt is made here to report the likelihood of individual species occurring on particular islands using logit regression.

In the prediction of numbers of species on islands, analysis has been limited to the manageable number of islands for which there was at least one record on 1 October, 1995, that is for species regarded as being resident on the British mainland. The number of species is found by the expression:

$$\log S = -0.093 + 0.775 \log \mathbf{FS1} - 0.140 \log \mathbf{I1}+1 + 0.086 \log \mathbf{A}.$$

The predicted number of species is given in Table 6, together with 'high' and 'low' estimates defined by the 95% confidence limits around the regression line. Values of **FS1**, **I1** and **A** are also given for each island. When considering the predicted numbers, it is well to remember that some 30% of the variance in island species' richness has not been accounted for by faunal source, island isolation, island area and latitude. Also important to bear in mind is that the selection of the 73 islands on which predictions are based, using arbitrary criteria such as the

Table 6. The predicted number of species for partially surveyed islands off the British and Irish mainlands.

REGION	ISLAND	GRID	FS1	A	I1	PREDICTED S	S_{low}	S_{high}
Channel Islands	Jethou	WV38	67	19	40	16	14.3	17.9
South-west England	Looe	SX25	38	9	0.7	15	13.4	17.0
	St. Michael's Mount	SW52	36	9	0.5	15	13.0	16.6
Isles of Scilly	White	SV91	36	15	43.6	10	7.9	11.4
	St. Helen's	SV81	36	20	46	10	8.0	11.6
	Tean	SV91	36	16	45.1	10	7.9	11.4
	Samson	SV81	36	39	49.2	10	8.5	12.1
	Great Ganilly	SV91	36	13	42	10	7.8	11.3
	Great Arthur	SV91	36	8	43	9	7.4	10.9
	Gugh	SV80	36	38	49	10	8.5	12.0
	Round	SV91	36	4	46	9	6.7	10.3
	Menawethan	SV91	36	3	41.5	8	6.7	10.2
	Great Innisvoules	SV91	36	2	41.5	8	6.4	10.0
Bristol Channel	Flatholm	SS26	37	23	4.3	14	12.0	15.5
South Wales	Penrhyn-Gwyr´	SS38	36	15	0.4	16	13.9	17.4
	Caldey	SS19	37	188	1.1	19	17.0	20.5
Isle of Man	Chicken Rock	SC16	21	1	57	5	3.0	6.7
South-west Scotland/ Firth of Clyde	Lady Isle	NS22	21	5	3.7	8	6.1	9.7
	Horse	NS24	21	4	1.3	9	6.8	10.4
	Inchmarnock	NS05	22	324	3.7	12	10.0	13.5
	Holy	NS02	21	246	20	9	7.2	10.7
	Little Cumbrae	NS15	20	273	2.4	11	9.5	13.0
	Great Cumbrae	NS15	20	2072	1.7	14	12.0	15.6
	Pladda	NS01	24	11	25.5	7	5.6	9.1
	Sanda	NR70	18	179	2.6	10	8.1	11.7
	Glunimore	NR70	18	2	3.5	7	4.7	8.3
	Sheep	NR70	18	8	2.4	8	5.9	9.4
	Davaar	NR71	18	56	0.6	10	8.3	11.8
	Sgat Mor	NR96	22	2	0.4	9	7.2	10.8
	Glas Eilean	NR98	22	1	0.25	9	6.8	10.4
	Eilean Aoghainn	NR99	22	2	0.9	9	6.8	10.4
	Burnt Island	NS07	22	2	0.2	9	7.4	11.0
Inner Hebrides/ Strathclyde	Cara	NR64	20	63	4.5	9	7.5	11.0
	Oronsay	NR38	23	187	32	9	7.1	10.6
	Scarba	NM36	23	1831	4.5	14	12.0	15.5
	Garvellachs	NM60+NM61	23	175	9	10	8.6	12.1
	Luing	NM70	23	1357	1.3	15	13.4	16.9
	Lunga	NM70	23	194	6.7	11	9.1	12.6
	Easdale	NM71	23	19	3.8	9	7.7	11.3
	Lismore	NM73	22	2500	1	16	14.0	17.5
	Treshnish Isles (mean)	NM23	22	15	31	7	5.1	8.7
	Soa	NM21	22	13	45	6	4.7	8.3
	Ulva	NM43	22	2300	16.5	12	9.8	13.3
	Staffa	NM33	22	28	28.8	7	5.6	9.1
Inner Hebrides/ Highland	Muck	NM47	21	526	8.2	11	9.0	12.5
	Eilean nan Each	NM38	21	31	11.7	8	6.3	9.8

Table 6 cont.

REGION	ISLAND	GRID	FS1	A	I1	PREDICTED		
						S	S_{low}	S_{high}
	Eigg	NM48	21	2973	11.8	12	10.1	13.6
	Eilean Mór	NG63	18	193	1.6	10	8.7	12.2
	Eilean Meadonach	NG63	18	54	2.1	9	7.3	10.9
	Sanday	NG20	18	191	37.3	7	5.4	8.9
	Heisker	NM19	21	19	46.5	6	4.6	8.2
	Soay	NG41	18	1009	25.5	9	6.9	10.4
	Wiay	NG23	18	178	24	8	5.8	9.3
	Longay	NG63	18	51	7.1	8	6.2	9.7
	Pabay	NG62	18	133	7	9	6.9	10.4
	Fladday	NG55	17	137	8.8	8	6.3	9.8
	South Rona	NG65	17	976	6.4	10	8.1	11.7
North-west coast of Scotland	Longa	NG77	17	135	1	10	8.3	11.8
	Eilean Furadh Mor	NG89	17	23	0.5	9	7.2	10.8
	Summer Islands	NB90	15	620	1	10	8.6	12.2
	Tanera Beg	NB90	15	86	2.6	8	6.3	9.9
	Tanera Mor	NB90	15	295	1	10	8.0	11.5
	Preist	NB90	17	126	4.9	9	6.8	10.3
	Carn nan Sgeir	NC00	15	4	2.3	6	4.5	8.1
	Handa	NC14	15	321	0.4	10	8.5	12.1
Outer Hebrides	Berneray	NL57	20	202	86.6	7	5.2	8.7
	Mingulay	NL58	20	647	87.6	8	5.9	9.4
	Pabbay	NL68	20	245	83.4	7	5.3	8.9
	Sandray	NL69	20	388	80.7	7	5.6	9.2
	Vatersay	NL69	20	953	82.5	8	6.2	9.8
	Muldoanich	NL69	20	79	78.5	6	4.7	8.3
	Flodday	NL69	20	27	84.1	6	4.1	7.7
	Uinessan	NL69	20	5	82.4	5	3.3	6.9
	Gighay	NF70	20	88	75.8	7	4.8	8.4
	Hellisay	NF70	20	137	75.7	7	5.1	8.6
	Fuday	NF70	20	224	80	7	5.3	8.9
	Fiaray	NF71	20	35	85.5	6	4.2	7.8
	Eriskay	NF70	20	1360	75.1	8	6.6	10.1
	Calavay	NF85	20	4	78.6	5	3.2	6.8
	Wiay	NF84	17	344	80.4	6	4.7	8.2
	Monach Isles	NF66	17	353	104.7	6	4.5	8.0
	Ronay	NF85	17	490	78.6	7	4.9	8.5
	Grimsay	NF85	17	591	81.4	7	5.0	8.5
	Baleshare	NF75	17	806	89.6	7	5.1	8.6
	Berneray	NF98	17	1165	80.1	7	5.4	9.0
	Pabbay	NF88	17	732	83.6	7	5.1	8.6
	Shillay	NF89	17	41	85.5	5	3.6	7.1
	Ensay	NF98	17	250	75.8	6	4.6	8.1
	Killegray	NF98	17	156	75.8	6	4.3	7.9
	Taransay	NF99	17	1375	71.7	7	5.6	9.2
	Scarp	NA91	17	1002	78.7	7	5.3	8.9
	Scotasay	NB19	17	49	56.5	6	4.0	7.5
	Great Bernera	NB13	17	2012	74.5	8	5.8	9.4
	Little Bernera	NB14	17	113	86.8	6	4.0	7.6
	Shiant Islands	NG49	17	140	30.6	7	5.1	8.6
	St Kilda (Hirta)	NF09	17	647	163.2	6	4.4	8.0
	North Rona	HW83	15	120	80	5	3.6	7.2

Table 6 cont.

REGION	ISLAND	GRID	FS1	A	I1	PREDICTED		
						S	S_{low}	S_{high}
Orkney Isles	Burray	ND49	13	900	24.4	7	5.0	8.5
	Graemsay	HY20	13	382	28.8	6	4.3	7.9
	Shapinsay	HY51	13	2688	44	7	5.0	8.6
	Rousay	HY42	13	4523	54.1	7	5.2	8.7
	Westray	HY43	13	4480	67.3	7	4.9	8.5
	Sanday	HY63	13	2798	65.3	6	4.7	8.2
Shetland Isles	Bressay	HU53	13	3100	193.5	6	3.8	7.4
	Noss	HU53	13	400	200.3	5	2.9	6.5
	Whalsay	HU56	13	2590	214.8	5	3.7	7.2
	Fetlar	HU68	13	4400	241.6	6	3.8	7.4
	Papa Stour	HU15	13	1942	197.7	5	3.6	7.2
	Mousa	HU42	13	159	180.2	4	2.6	6.2
	Outer Skerries	HU67	13	600	231.3	5	3.0	6.6
	Unst	HU95	13	12090	251.9	6	4.3	7.9
	Uyea	HU69	13	218	251.3	4	2.5	6.1
East Scotland/ Firth of Forth	Bass Rock	NT68	18	6	2.1	8	5.8	9.4
	Fidra	NT58	18	5	0.7	8	6.3	9.9
	Inchkeith	NT28	18	22	3.9	8	6.1	9.7
	Inchmickery	NT28	17	12	3	7	5.6	9.2
Northumberland	West Wideopens	NU23	18	3	2.8	7	5.1	8.7
	Longstone	NU24	18	2	6.6	6	4.3	7.9
	South Wamses	NU25	18	1	5.4	6	4.0	7.7
	Brownsman	NU26	18	5	5.1	7	5.0	8.6
	Staple	NU27	18	4	4.9	7	4.9	8.5
	North Wamses	NU28	18	2	5.5	6	4.4	8.0
	Northern Hares	NU29	18	1	6.8	6	3.9	7.5
Ireland	Garinish	0095	24	4	0.15	10	8.7	12.3
	Dursey	0044	24	558	0.35	16	13.9	17.4
	Bear	0074	24	1791	2.5	15	13.4	16.9
	Scarrif	0045	24	136	5	11	9.5	13.0
	Puffin	0036	24	47	0.2	13	11.1	14.6
	Valencia	0047	24	1917	0.25	18	15.8	19.3
	Great Blasket	0029	24	434	1.8	14	12.1	15.6
	Inishman	0290	26	924	12.7	13	10.8	14.3
	Inisheer	0290	26	563	7.5	13	11.1	14.6
	Omey	0255	25	207	0.4	15	13.0	16.5
	Gorumna	0282	25	1349	0.7	17	15.1	18.6
	Inishbofin	0256	24	640	5.4	13	11.0	14.5
	Clare	0268	22	1628	5	13	11.3	14.8
	Aran	1461	24	3089	3	16	13.8	17.3
	Cruit	1472	24	291	0.1	15	13.5	17.0
	Tory	1484	24	331	10	11	9.4	12.9

FS1, size of nearest mainland source; A, area (ha); I1, isolation from nearest mainland source (km).

Treshnish Island areas: Bac Beog, 7.9; Bac Mor, 22.4; Lunga, 64.8; Fladda, 22.6; Cairn na Burgh Beg, 1.7; Cairn na Burgh Mor, 3.6; Sgeir a Chaisteil, 4.9; Sgeir an Eirionnaich, 4.2; Sgeir an Fheoir, 0.8.

Predicted number of species: $\log S = -0.093 + 0.775 \log(FS1) - 0.140 \log(I+1) + 0.086 \log(A)$.

S_{low} and S_{high} are the 95% confidence limits for the predicted number of species.

Table 6 cont.

Stepwise multiple regressions of numbers of butterfly species on 73 islands against four geographic variables.

a) S on A, I1, FS1, L2 (all normalised)

First variable		beta	R^2	R^2_{adj}	$F_{(171)}$	P
	FS1	0.642	0.412	0.404	49.706	<0.001
FS1	Intercept	0.199	t = 0.588			NS
	FS1 slope	0.731	t = 7.05			<0.001

Second variable		beta	R^2	R^2_{adj}	$F_{(270)}$	P
	A	0.403	0.563	0.551	45.178	<0.001
A	FS1	0.745				
	Intercept	–0.719	t = 2.065			<0.05
	FS1 slope	0.848	t = 9.118			<0.001
	A slope	0.075	t = 4.932			<0.001

Third variable		beta	R^2	R^2_{adj}	$F_{(369)}$	P
	I1	–0.397	0.712	0.696	56.746	<0.001
I1	FS1	0.681				
	A	0.461				
	Intercept	–0.214	t = 0.72			NS
	FS1 slope	0.775	t = 10.056			<0.001
	A slope	0.086	t = 6.819			<0.001
	I1 slope	–0.14	t = 5.953			<0.001

Fourth variable		beta	R^2	R^2_{adj}	$F_{(468)}$	P
	L2	–0.228	0.718	0.701	43.255	<0.001
L2	FS1	0.478				
NON-	A	0.493				
SIGNIFICANT	I1	–0.396				
	Intercept	8.958	t = 1.2			NS
	FS1 slope	0.544	t = 2.683			<0.01
	A slope	0.092	t = 6.829			<0.001
	I1 slope	–0.14	t = 5.964			<0.001
	L2 slope	–2.119	t = 1.23			NS

occurrence of certain long-distance migrants, is no guarantee that the island has been properly surveyed nor that the islands have been surveyed with the same degree of intensity. If anything, our predictions for islands will underscore the numbers of species occurring on them. In this respect, the upper and lower bounds provided by the 95% confidence limits give a reasonable estimate of how many species can be expected to be recorded in different circumstances. It does not account for extreme conditions. For example, islands with base rich and acid geological substrates, including much range in altitude and variation in slope angles, may expect to have a wide variety of habitats and greater species' richness. Conversely, islands densely colonized over their entire surface by nesting seabirds

Table 6 cont.

b) S on A, I2, FS2, L1 (all normalised)

First variable		beta	R^2	R^2_{adj}	$F_{(171)}$	*P*
	FS2	0.642	0.62	0.615	115.874	<0.001
FS2	Intercept	0.417	t = 2.05			<0.05
	slope	0.716	t = 10.765			<0.001

Second variable		beta	R^2	R^2_{adj}	$F_{(270)}$	*P*
	A	0.355	0.742	0.735	100.782	<0.001
A	FS2	0.847				
	Intercept	–0.219	t = 1.087			NS
	FS2 slope	0.77	t = 13.76			<0.001
	A slope	0.066	t = 5.76			<0.001

Third variable		beta	R^2	R^2_{adj}	$F_{(369)}$	*P*
	L1	–0.315	0.787	0.777	84.832	<0.001
L1	FS2	0.638				
	A	0.431				
	Intercept	11.408	t = 3.715			NS
	FS2 slope	0.58	t = 8.103			<0.001
	A slope	0.08	t = 7.193			<0.001
	L1 slope	–2.792	t = 3.793			<0.001

Fourth variable		beta	R^2	R^2_{adj}	$F_{(468)}$	*P*
	I2	–0.114	0.799	0.787	67.62	<0.001
I2	FS2	0.633				
	A	0.447				
	L1	–0.308				
	Intercept	11.23	t = 3.739			<0.001
	FS2 slope	0.576	t = 8.216			<0.001
	A slope	0.083	t = 7.572			<0.001
	L1 slope	–2.728	t = 3.787			<0.001
	I2 slope	–0.057	t = 2.049			<0.05

and extensively covered by their detritus, as possibly on Moelfre Island off Anglesey, are likely to have an impoverished butterfly fauna.

One aspect of the present data on islands is that it gives cumulative records over a number of decades. As such, and bearing in mind what was said about turnover rates in species' numbers, as well as data on the ability of butterflies to cross sea surfaces (see section I.7), islands close to mainland sources would be expected to have a cumulative number of species over decades that approaches an asymptote the limit for which is determined by the number of species at the nearest mainland faunal source. However, the number of species should remain much the same from one year to the next. This is largely the case for Hilbre Island off the Wirral peninsula in Cheshire; data for Hilbre suggest the occurrence of distinct events, cases of vagrants, colonization and extinction, as well as population fluctuations influenced by seasonal weather conditions (see Appendix 2).

7. MIGRATION RECORDS

Interpretation of butterfly records for the British islands, notably of the factors underlying species' richness, has been influenced inevitably by our perception of the ability of butterflies to migrate across open water and to colonize isolated habitats. Data on dispersal and migration are particularly difficult to obtain, but those which have accumulated over the last century would appear to evoke a duality of expectations. On the one hand, few working on butterflies have difficulty envisaging the distant and reversed migrations, *en masse*, of species such as *Vanessa atalanta*, *Cynthia cardui*, *Pieris brassicae*, *P. rapae* and *Colias croceus*; on the other hand, the ability of most resident species to cross less than 1 km of open countryside to occupy vacant habitats is often viewed with understandable sceptism. This dualism may largely be a legacy from Ford (1964). Whilst discussing the genetic homogeneity among *Maniola jurtina* populations on the three largest of the Isles of Scilly he had this to say of migration and gene flow: 'We have found that even a hundred yards of unsuitable territory is an almost complete barrier to this insect ...' (Ford, 1964: 57). He further emphasized the point regarding movement between habitat patches on Tean: 'One could see the butterflies setting out in numbers over the two 'lawns' from either end and, finding them continuously inhospitable, turning back as over the sea, after about ten yards; in the middle none was to be seen.' (Ford, 1964: 60).

Debate on the ability of resident butterflies to cross inhospitable terrain has been a long one. For example, it was rigorously contested in the 1930s regarding the capacity of *Eurodryas aurinia* to disperse and establish new colonies in southern Britain. Opinions differed even to its speed of flight (Campbell-Taylor, 1931a, b; Curtis, 1931, 1932; Wheeler, 1931a, b; Castle-Russell, 1932). In this exchange of views, in which there was a complete disregard for changing conditions, the field evidence supported the ability of the butterfly to disperse over large distances (Huggins, 1972; Horton, 1977). Prior to Ford (1957, 1964) evidence from other island groups also pointed to the ability of resident butterflies to cross the sea and to reach islands (Campbell, 1952). In recent years the controversy largely continues. Autecological research on butterfly populations, applying MRR, indicates very limited movement between adjacent populations (Thomas, 1983a; Thomas, 1985; Warren, 1987a), even for species having supposedly open population structures (Courtney, 1980; Dennis, 1986; Dennis & Bramley, 1985), whereas the use of other techniques, for instance direct tracking or ex-habitat observations, clearly indicates the capacity of many of our resident species to cross inhospitable areas lacking any resource (Baker, 1969). Different conclusions even emerge using the same technique on the same butterfly. Our own observations on *Maniola jurtina* (Shreeve, Dennis & Williams, 1995), again applying MRR but also from direct observations, contrast markedly with those of

Ford and his co-workers (Ford, 1964; Dowdeswell, 1981); we demonstrate the butterfly to cross, in numbers over distances of several hundred metres, a range of unsuitable habitats such as woods, expanses of mown grass, urban areas and open water. Furthermore, there is probably ample evidence that the majority of our butterfly fauna have the capacity to engage in dispersal (Shreeve, 1992, 1995).

Before advancing further with this discussion on insect movement, it is as well to appreciate what is meant by non-habitat and movements that are ex-habitat. Insect migration (movements between spatial units) is adapted to the distribution of resources, usually envisaged as occurring in a single habitat, and to spatial changes in habitat location, owing mainly to vegetation succession or to landscape processes (see Baker, 1978). Appropriate habitats (H) for a butterfly species necessarily comprise distinct sets of resources. Obvious and important ones are:

(i) mate-location and courtship sites (C);
(ii) oviposition sites (O);
(iii) hostplant resources (L);
(iv) nectar and adult-feeding resources (N);
(v) roost sites (R);
(vi) predator-escape sites (E);
(vii) overwintering sites (W).

A highly suitable habitat would occur where these resources completely overlap. In set notation, where each resource can be depicted by an envelope enclosing it on a map:

$$C \cup O \cup L \cup N \cup R \cup E \cup W = H \qquad \text{(Fig. 8a).}$$

Essentially, in this case, little movement would be necessary for an insect requiring these resources in succession. Even less movement is predicted for an insect in which all the resources involve the same medium. Thus, where:

$$C \Leftrightarrow O \Leftrightarrow L \Leftrightarrow N \Leftrightarrow R \Leftrightarrow E \Leftrightarrow W$$

then

$$H = C = O = L = N = R = E = W \qquad \text{(Fig. 8b).}$$

Suitable habitats for a species may still be found where these resources do not overlap but are contiguous, such that:

$$C \cap O \cap L \cap N \cap R \cap E \cap W = \emptyset$$

but

$$C \cup O \cup L \cup N \cup R \cup E \cup W = H \qquad \text{(Fig. 8c).}$$

In situations where specific resources are contiguous but do not overlap, individuals would occasionally need to fly through resource zones that fail to satisfy an immediate stimulus. But, where these are not contiguous, then individuals would need to fly over ground lacking in any resource at all (Fig. 8d). There are, of course, resource-vacant areas even within zones where all the

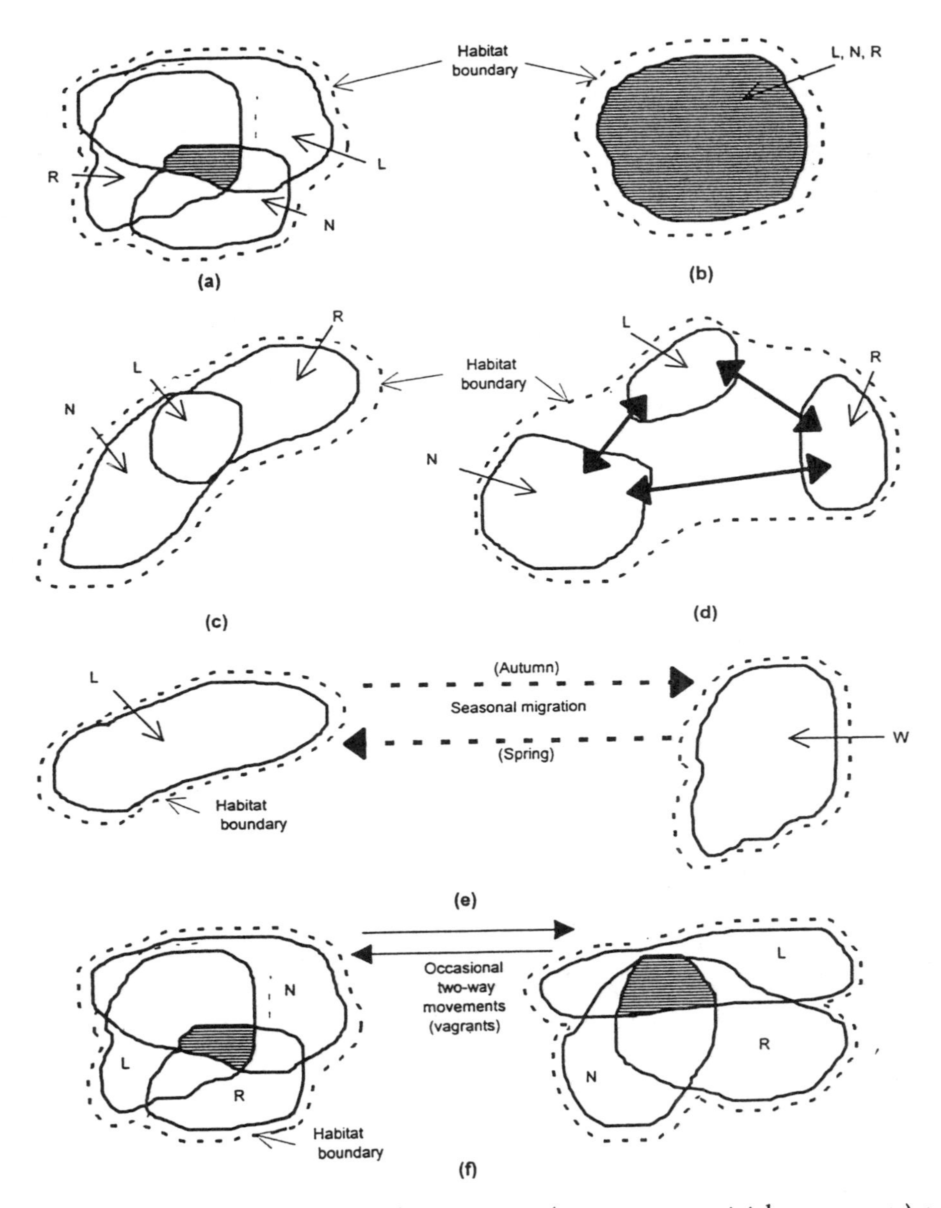

Fig. 8. The relationship of individual movements (i.e., vagrants, trivial movements) to habitat structure. For simplicity, a maximum of three components of habitat structure are illustrated in each diagram. N, nectar resource; L, larval resource; R, roost sites; W, overwintering sites. Habitat resource classification: (a) intersection and union; (b) equivalence and equality; (c) contiguous union; (d) disjointed union linked by daily trivial flights; (e) disjointed non-union linked by seasonal migration; (f) metapopulation of two habitats linked by random migration. Shading, resources overlap and habitat core. See text for detail.

necessary resources overlap. These may be considered as part of a habitat, as insects continually move through this space during daily trivial flights from one resource element (e.g., a shoot or patch) to another. Non-habitat is defined, for the purpose of this work, as surfaces lying outside the path of daily trivial flights and lacking in any resource for enhancing somatic maintenance or reproduction. Thus, ground, lacking resources, which needs to be flown over during passage between seasonal resources is regarded as non-habitat and the flights as occurring ex-habitat (Fig. 8e). Such flights together with those that occur randomly out of or between habitats (Fig. 8f) are considered to be undertaken by vagrants. It should be noted that many metapopulation models assume equivalence and equality of resources as shown in Fig. 8b (e.g., Hanski, 1994; Hanski & Thomas, 1994). The physical gaps between resources within habitats and between habitats form the basis of, and select for, movements within insect populations (Baker, 1978).

One, as yet, insuperable problem in the study of movements is that all techniques are in some way limited or flawed. MRR may well influence the capacity or readiness of insects to move, and in different circumstances could increase or decrease ex-habitat dispersal (Morton, 1982). Using MRR, sampling problems also occur as the area of study increases, making it difficult to pick up long-distance dispersing individuals representing the tail of the leptokurtic (peaked) and positively-skewed distribution curve of movements. Direct tracking can also disrupt behaviour and is limited by our capacity to follow the insects in flight. New techniques, using pollen or chemical markers (Courtney, Hill & Westerman, 1982; Dempster, Lakhani & Coward, 1986) face other problems and are not reliable. New approaches are needed (Roland *et al.*, 1996). Here, we review data from a number of sources; these suggest, for all their limitations, that our resident species have a considerable capacity to migrate across unsuitable terrain and confirm observations made 25 years ago by Baker (1969, 1978). But, we would draw an important distinction between migration and colonization; the ability of an insect to cross landscape is not the same as its ability to found new colonies (Shreeve, 1992, 1995). Not only must the conditions be suitable for the colonist but establishment may depend on a few, even on one, gravid females.

Several sources of data are available on butterfly movements that would suggest that most, if not all, species are capable of crossing resource-vacant terrain including sea barriers. These comprise:

(i) sea records from light vessels, ships, boats and shoreline observations;
(ii) island vagrant records for species that lack resources for reproduction (i.e., hostplants and habitat components) or survival (i.e., for overwintering);
(iii) observations of ex-habitat vagrants;
(iv) suburban garden records;
(v) city central business district (CBD) records;
(vi) long distance migrations typically with seasonal reversals in direction;
(vii) mass movements often involving several species flying in one direction;
(viii) expansions in geographical range.

Data from these sources are scored under seven variable terms to allow comparison with earlier designations of mobility from autecological surveys (i.e., M_1, M_2; Thomas, 1984; Pollard & Yates, 1993) (Table 7):

- sea records (**SR**);
- ex-habitat vagrants (**VA**);
- suburban garden records (**SG**);
- CBD records (**UC**);
- range expansions (**RE**);
- frequent long distance migrations (**LDM**);
- mass movements (**MM**).

From these variables, a mobility index (**MI**) has been calculated, which is simply the sum of the binary states for each variable.

These data contest the often quoted general finding of MRR techniques applied to single colonies or limited metapopulations, that is, few butterfly species have the capacity to migrate to offshore islands. Of the 60 species that are currently regarded as British butterflies, 47 are described as having 'closed' populations (Thomas, 1984: Table 33.2 to include *Gonepteryx rhamni*), 32 are classed as 'sedentary', and 50 as not being 'wide-ranging' (Pollard & Yates, 1993). The implication is that even short stretches of intervening ground deficient in resources for species (i.e., egg-laying, nectaring, roosting, mate-location etc.) will

Table 7. The population structure and mobility of 57 butterfly species resident on the British and Irish islands.

SPECIES	M_1	M_2	SR	VA	SG	UC	RE	LDM	MM	MI
1 *C. palaemon*	C	S	0	0	0	0	0	0	0	0
2 *T. sylvestris*	C	S	1	1	1	1	1	0	0	5
3 *T. lineola*	C	S	0	1	1	1	1	0	0	4
4 *T. acteon*	C	S	0	0	0	0	0	0	0	0
5 *H. comma*	C	I	0	1	0	0	1	0	0	2
6 *O. venata*	C	S	1	1	1	1	1	0	0	5
7 *E. tages*	C	S	0	1	1	0	0	0	0	2
8 *P. malvae*	C	S	0	1	1	0	0	0	0	2
9 *P. machaon*	C	I	2	1	1	0	0	0	1	4
10 *L. sinapis*	C	I	0	1	0	0	1	0	0	2
11 *G. rhamni*	O	W	1	1	1	1	0	0	0	4
12 *P. brassicae*	O	W	2	1	1	1	?	1	0	5
13 *P. rapae*	O	W	2	1	1	1	?	1	0	5
14 *P. napi*	O	I	2	1	1	1	?	1	0	5
15 *A. cardamines*	O	I	1	1	1	1	1	0	0	5
16 *C. rubi*	C	S	0	1	0	1	0	0	0	2
17 *T. betulae*	C	I	0	0	0	0	0	0	0	0
18 *Q. quercus*	C	S	0	1	1	0	0	0	1	3
19 *S. w-album*	C	S	0	1	1	0	0	0	0	2
20 *S. pruni*	C	S	0	0	0	0	0	0	0	0
21 *L. phlaeas*	C	I	2	1	1	1	1	0	1	6
22 *L. dispar*	C	S	0	0	0	0	0	0	0	0
23 *C. minimus*	C	S	0	1	1	0	0	0	0	2
24 *P. argus*	C	S	0	1	0	0	1	0	0	2
25 *A. agestis*	C	I	0	1	0	0	0	0	0	1

Table 7 cont.

SPECIES	M_1	M_2	SR	VA	SG	UC	RE	LDM	MM	MI
26 *A. artaxerxes*	C	S	0	1	1	0	0	0	0	2
27 *P. icarus*	C	I	2	1	1	1	1	0	1	6
28 *L. coridon*	C	S	0	1	1	0	0	0	0	2
29 *L. bellargus*	C	S	0	0	0	0	0	0	0	0
30 *C. argiolus*	O	W	1	1	1	1	1	0	0	5
31 *M. arion*	C	S	0	0	0	0	0	0	0	0
32 *H. lucina*	C	S	0	0	0	0	0	0	0	0
33 *L. camilla*	C	I	1	1	1	1	1	0	0	5
34 *A. iris*	C	I	0	1	0	0	0	0	0	1
35 *A. urticae*	O	W	2	1	1	1	?	1	0	5
36 *N. polychloros*	O	W	2	1	1	1	0	0	1	5
37 *I. io*	O	W	2	1	1	1	1	1	0	6
38 *P. c-album*	O	I	1	1	1	1	1	0	1	6
39 *B. selene*	C	S	1	1	0	0	0	0	0	2
40 *B. euphrosyne*	C	S	0	1	1	0	0	0	0	2
41 *A. adippe*	C	I	0	1	0	1	0	0	0	2
42 *A. aglaja*	C	I	1	1	1	1	1	0	0	5
43 *A. paphia*	C	I	1	1	0	1	0	0	1	4
44 *E. aurinia*	C	I	0	1	1	0	0	0	0	2
45 *M. cinxia*	C	I	0	0	0	0	0	0	0	0
46 *M. athalia*	C	S	0	0	0	0	0	0	0	0
47 *P. aegeria*	C	S	1	1	1	1	1	0	0	5
48 *L. megera*	C	?	2	1	1	1	1	0	0	5
49 *E. epiphron*	C	S	0	0	0	0	0	0	0	0
50 *E. aethiops*	C	S	0	1	0	0	0	0	0	1
51 *M. galathea*	C	S	0	1	1	0	0	0	0	2
52 *H. semele*	C	S	1	1	1	0	0	0	0	3
53 *P. tithonus*	C	S	0	1	1	1	1	0	0	4
54 *M. jurtina*	C	S	1	1	1	1	0	0	1	5
55 *A. hyperantus*	C	S	0	1	1	0	1	0	0	3
56 *C. pamphilus*	C	S	1	1	1	1	0	0	0	4
57 *C. tullia*	C	S	0	1	0	0	0	0	0	1

M_1: mobility (from Thomas, 1984); C, closed populations; O, open populations.
M_2: mobility (from Pollard & Yates, 1993); S, sedentary; I, intermediate; W, wide ranging
SR: sea-records from light vessels or documented records of sea crossings; 1, record(s); 2, sea crossings >10 km; 0, no records.
VA: vagrants recorded ex-habitat; 1, record(s); 0, no records.
SG: suburban garden records; 1, record(s); 0, no records.
UC: records from Inner London, Oxford, Birmingham and Sheffield centres; 1, record(s); 0, no records.
RE: documented range expansions; 1, evidence of expansion; 0, no evidence of range expansion.
LDM: long-distance movements; 1, frequent long-distance reversed migrations; 0, no records.
MM: documented mass movements; 1, numbers moving ex-habitat, often with other species; 0, no records.
MI: movement index. Sum of binary states for seven variables, **SR** to **MM**.

The data on movements and migration **(SR–MM)** are recorded as presence or absence of records irrespective of the number of records. Doubtful records are excluded and no scaling is given to the number of records in any category since none of these data has been systematically collected.

Note: *H. morpheus*, *C. croceus*, *V. atalanta* and *C. cardui* excluded.

isolate colonies of these species. Isolation is argued to be distance-dependent and to vary in magnitude for different species. For example, it has been argued, from MRR data on *Lysandra bellargus*, that colonies may be effectively isolated by as little as 100 m of unsuitable ground (Thomas, 1983a), whereas marked *Leptidea sinapis* have been found as much as 4 km away from their point of initial capture (Warren, Pollard & Bibby, 1986). Applying logistic regression to occupied and unoccupied habitat patches of different size and degree of isolation, attempts have been made to deduce the maximum natural single-step colonization distances for several species. In *Plebejus argus* this is credited to be 0.6–1.0 km, for *Hesperia comma* 8.65 km, for *Thymelicus acteon* 2.25 km, for *Mellicta athalia* 0.65–2.5 km, and *Satyrium pruni* 1.4 km (Thomas & Jones, 1993; Thomas, Thomas & Warren, 1992). It is important to realize that these observations mainly refer to the rarer elements of the British fauna measured over limited periods of time; as for these species conditions may be distinctly marginal throughout Britain, estimates of isolation made over short periods may underscore migration (see section I.8 and 10). However, the conclusions from these estimates are more realistic and coincide with findings in the present work: 'a lack of suitable habitat patches for a few to tens of kilometres would prevent the spread of species from regions of occupied patches to those which are vacant' (Thomas *et al*., 1992).

Data on sea crossings derive from two sources:

(i) at-sea records;
(ii) island records for vagrants lacking resources for reproduction and survival.

Examination of the entomological literature discloses that there is good evidence for some 27 of the 60 recorded resident species having made sea crossings; twelve are regarded as having 'closed' populations and seven as being sedentary, representing 44% and 22% of species in those categories respectively (Table 7). These figures may significantly underscore such movements as the data have been accumulated from casual records and have not been compiled from any systematic survey. For 16 of the 27 species, the minimum distance travelled across the sea exceeds 10 km. This is the distance to the nearest landmass and does not take the direction of movement into consideration. Although light vessel records often give flight direction, this is for a single point in time and, though suggestive of direction of origin, without additional data cannot simply be assumed to indicate the direction of movement for the whole period prior to the time of observation. Some resident species have been found much further out to sea, as in the case of *Lasiommata megera* 48 km out on Outer Dowsing light vessel off Spurn Head (Dannreuther, 1933). Out at sea, butterflies, even resident species, have been found flying against the wind (e.g., *Maniola jurtina* and *Lycaena phlaeas* at Sovereign light vessel on 1 July 1933 and 29 August 1933 respectively; *Coenonympha pamphilus* at Gorleston Pier, 27–31 August 1933) as well as with it (e.g., *Lasiommata megera* at Outer Dowsing light vessel, 20 August 1933) (Dannreuther, 1933). Some unusual behaviour has been observed at sea, such as the 'spiralling cloud' of 30 or more *Aglais urticae* observed midway between Troon and Lady Isle (Gibson, 1982d). One significant problem for overseas flight, compared to dispersal over land, is the lack of places for insects to rest when exhausted or during the night. Even so, several butterflies have been recorded as flying in the dark (e.g., *Pieris napi*: Heslop Harrison, 1940c; *Inachis io*: Frazer, 1939; *Satyrium w-album*: Kett, 1993; Bristow, 1994); *Vanessa atalanta*, *Cynthia cardui* and *Aglais urticae* (e.g., Tremewan, 1953) have frequently been recorded doing so, even in storm conditions (Sullivan, 1946). Records also exist of butterflies landing on water and taking off again (e.g., *Pieris rapae*: Dowdeswell & Ford, 1948; *Quercusia quercus*: Holloway, 1980; *Coenonympha pamphilus*: Shreeve, pers. obs.), even repeatedly and sailing along in rough weather (e.g., *M. jurtina*: Heslop Harrison, 1946a). Observations exist which suggest that butterflies can also take sustenance from water bodies when they are in flight (e.g., *Vanessa atalanta*: G. Bennett, pers. comm.). What initiates these movements over water is not specifically known, but butterflies have been observed leaving the shoreline and heading out to sea (e.g., *Hipparchia semele*: Campbell, 1952; *Ladoga camilla*: Birtley *in* Dannreuther, 1935; *Argynnis aglaja*: Heslop Harrison, 1939).

Numerous records exist of vagrants occurring on islands without resources for reproduction or survival over the winter months. Many have been recorded for the British mainland and Ireland (e.g., *Iphiclides podalirius*: Wilkinson, 1975, 1982; see Emmet & Heath, 1989), but also for the Isle of Wight (e.g., *Papilio machaon*, *Aporia crataegi*, *Cynthia virginiensis* and *Argynnis lathonia*: Fearnehough, 1972) and the Channel Islands (Long, 1970, 1987). Some familiar immigrants have been found on several of the smaller islands (e.g., *Danaus plexippus* on Skomer, St Agnes and Tresco; *Lampides boeticus* on Hayling and Isle of Wight; *Nymphalis antiopa* on Foula, Great Saltee and Lindisfarne) and these are believed in the main to have flown to the islands, much as those species which frequently engage in long-distance and reversed migrations. However, numerous exotics have been recorded that could have been accidently or deliberately introduced (e.g., *Melitaea didyma*, *Lasiommata maera*, *Gonepteryx cleopatra*; see Emmet & Heath, 1989). Occasionally, species that are resident on the British mainland are also observed on islands where clearly neither breeding resources nor overwintering resources exist. Some care is needed in interpreting observations of alleged vagrants as hostplant patches may have restricted distributions on islands and be missed by recorders. Obvious examples of vagrants on islands are *Argynnis paphia* on Bardsey and Inishtrahull (Darlington, 1954; Rippey, pers. comm.), *Papilio machaon* at Voe, North Mainland in the Shetlands (Pennington, pers. comm.; Baldwin, 1995) and *Polygonia c-album* on Lewis & Harris (Trevor, 1994). In each case the islands lack conditions for their continued survival. However, a fundamental problem with such records is that it is not always clear, without further information, whether these acts of dispersal are natural or assisted. In the case of *P. c-album* and *I. io* on Lewis and Harris, they may well have come over with timber from Denmark (Trevor, 1994); the records of *A. paphia*, if accurate, are more likely to be the result of voluntary displacement, as the islands are not involved in trade. Several species have been recorded at Inishtrahull light vessel flying in from the south (e.g., *Pararge aegeria*, *Hipparchia semele*: Dannreuther, 1939). Although accidental and deliberate introductions may occur frequently, the many examples of vagrants on islands clearly demonstrate the capability of many species to cross sea barriers.

Other indicators of potential for dispersal and migration are ex-habitat terrestrial records, records from suburban gardens and urban centres and data on range expansions. Altogether, vagrant individuals have been observed for 48 British butterfly species and some 20 species are known to engage in linear displacement frequently or very frequently (Baker, 1969; Table 7). Thirty-five species classed as having 'closed' populations and 23 species classed as being sedentary have been observed as vagrants, 74% and 72% of those categories respectively. Individual butterflies are occasionally found many miles from known colonies (e.g., *Eurodryas aurinia*: Huggins, 1972; Horton, 1977; *Lysandra coridon* and *Melanargia galathea*: Allan, 1949; *Argynnis aglaja*: Grimwood, 1965). Even butterflies supposedly epitomizing species having 'closed' populations, such as *Cupido minimus* and *Plebejus argus*, have been found as much as 1–2 km from their known colonies (Horton, 1977; Shreeve, pers. obs.; Dennis & Bardell, 1996).

Eight species have the ability to undergo frequent long distance migrations reversed in direction with the seasons (Table 7). For a further eight, there is evidence for long-distance movements or even of mass movements, with numbers of individuals belonging to different species moving over open countryside, ex-habitat, in the same direction. Most of these mixed-species mass movements involve the well-known long-distance migrants (e.g., *Vanessa atalanta*, *Cynthia cardui*, *Colias croceus*, *Aglais urticae*, *Pieris brassicae* and *P. rapae*), but residents unknown for long-distance migrations, some regarded as having 'closed' populations, are occasionally involved. One mass movement in Kincardineshire during July 1910 included *Argynnis paphia* and *Polyommatus icarus* and travelled up the Dee valley on a south-east wind from the coast (Fenton, 1948). Another mass movement of *E. aurinia*, but including other species, was described for the Salisbury Plain by Horton (1977). Holloway (1980) provided an account of a further movement down the Tillingbourne valley, Surrey, in the same year, 1976, of *Quercusia quercus*. Buckstone (1938) reported *Polygonia c-album* flying with *Vanessa atalanta* while Dannreuther (1933) described a movement of *Maniola jurtina* and *Lycaena phlaeas* with *P. brassicae*, *P. rapae*, *P. napi*, *Inachis io* and *A. urticae* for 32 km inland from the East Anglian coast during 20-30 July 1933, the butterflies crossing all obstacles in their path. Smith (1993) recorded 'several hundred' *Celastrina argiolus* at Spurn Head in 1992, thought to have migrated from Humberside to the south. Occasionally, these mixed-species migrations are observed to cross open sea, as in the swarm of pierids, nymphalids and lycaenids passing Tuskar lighthouse off Co. Wexford on 28 August 1957 (French, 1958).

Gardens, urban centres and the central business districts of cities also provide good evidence that butterflies can cross substantial obstacles, for at least two reasons:

(i) resources for reproduction are absent at the location where records are made, either in the garden or in the CBD, thus the species is recorded ex-habitat;
(ii) extensive areas lacking in any kind of resource have failed to deter migration, in which case the species must have migrated ex-habitat.

Thirty-eight species have been observed in suburban gardens, 25 (53% of the category) classed as 'closed' in population structure and 18 (56% of the category) classed as being sedentary (Table 7). For many of these species (e.g., *Pyrgus malvae*, *Cupido minimus*, *Aricia artaxerxes*, *Lysandra coridon*, *Eurodryas aurinia*) gardens are not appropriate habitats. Records for the CBDs of London, Sheffield, Birmingham and Oxford provide a more stringent measure of the capacity to migrate. Twenty-eight species have been recorded in the CBDs of these cities, 15 (32% of the category) classed as 'closed' in population structure and 8 (25% of the category) as being sedentary (Table 7). Where appropriate habitats have occurred for these species in the urban centres, these are typically isolated by unsuitable terrain from typical biotopes in which they are generally found (Owen, 1949, 1951) though railway embankments and cuttings, river and canal banks can ensure extensive penetration into city areas (McLeod, 1972). Clearly, both garden and CBD records indicate the capacity of most butterflies to migrate across inhospitable landscapes; as indicators of such movements the study of minimum distances travelled to neighbouring habitats should prove very instructive.

There is some indication that all the movements described above are periodic and vary with conditions (see section I.10). No doubt, they lie at the root of range 'expansions' that have been observed during the last century. In many ways, range expansions offer less convincing evidence for migration, as records for new localities for species may simply be the result of their having been overlooked in the past. However, not only is there good evidence for range expansions in recent years (Dennis, 1977, 1993; McAllister, 1993; Ellis, 1994) but that this has involved migration across extensive areas of unsuitable terrain, distances well in excess of those typically studied using MRR techniques, and the colonization of new habitats (Hardy, Hind & Dennis, 1993). Some 14 species are known to have expanded their ranges since 1945 and at least nine since 1988 (Pollard, Moss & Yates, 1995; Table 7). The latter include *Thymelicus sylvestris*, *Ochlodes venata*, *Anthocharis cardamines*, *Inachis io*, *Polygonia c-album*, *Pararge aegeria*, *Lasiommata megera*, *Pyronia tithonus* and *Aphantopus hyperantus*. *Celastrina argiolus* has also undergone a period of range expansion during this period (Hardy, Hind & Dennis, 1993; Pollard & Yates, 1993).

These data on movements suggest that the terms 'closed' and 'open' are inappropriate for describing population structure in species. First, it infers that species can be placed in one of two states based on data for inter-colony mobility. Second, the way the terms are used implies that the status of a species is 'fixed'; it either has a 'closed' or 'open' population structure. The three state categories of 'sedentary', 'intermediate' and 'wide-ranging' (Pollard & Yates, 1993) offer a more suitable alternative. However, there is still the problem that the terminology is based on inadequate data for butterfly movements. More important, it does not address the issue of spatial and temporal variation in mobility. In reality, what is being measured or described is not mobility but the type of metapopulation structure*; and this may change with time. Harrison (1991) described several categories (i.e., non-equilibrium; Boorman-Levitt; Levins; patchy population) that may form useful substitutes for the 'open' versus 'closed' terminology. Each

* A metapopulation can be defined as a system of population units within a landscape that is potentially linked by individual movements. Over time, local population extinction may occur, as well as local colonization, but the system of linked population units will persist at the regional scale as long as the fraction of occupied patches >0.

carries strong inferences about movement, about the frequency vagrants are likely to be observed, their variability over space and time and how they are selected for and against. Thus, a Levins metapopulation structure will appear to be 'closed', in comparison to a 'patchy population', for at least one reason; the local population size will often be smaller, so fewer individuals (not necessarily proportionally fewer) will disperse and fewer will be seen doing so. There is another reason that is based on dispersal being subject to selection as any other individual trait. As resource patches decline in number and area, so selection will increase against individuals dispersing, and the regional population of the species may experience a reduction in dispersal (Dennis, 1982b). It has also been observed that different metapopulation structures exist for the same species in different but not distant locations and at different times (e.g., *Pieris napi*, *Maniola jurtina*: Dennis, pers. obs.).

8. ECOLOGICAL BASIS FOR ISLAND BUTTERFLIES

In section I.3, an assessment of the contribution of factors to island species' richness was made. This analytic technique excludes consideration of species' associations on islands and at faunal sources. A relationship between the two is suggested in the ordination of similarities among islands based on their faunal assemblages; also, to some extent in the correlation between Jaccard similarities for actual data on island species' assemblages and those modelled using random inclusion of species from sets of proximate 'mainland' faunal sources (see section I.4 and Table 5). However, more direct measures of this relationship between islands and their sources can be made. If the faunal sources have any relevance at all for the make-up of island faunas, then the incidence of species on islands should correspond largely to their incidence at the nearest sources, and in turn to their geographical ranges (viz., latitudinal extent) and to their distributions. Furthermore, the ecological influences underlying this relationship can be explored by focusing on the ecological characteristics of the butterfly species occupying the islands. The basic question that can be asked is whether species found frequently on islands have greater propensity to migrate to, colonize and persist on islands than those that occur infrequently or not at all; that is, significant correlations should exist between the incidence of species on islands and relevant ecological parameters.

Geographical and ecological data on the butterflies are abstracted from Dennis (1993; see Table 8). The variables include:

- the incidence of species on islands (**IR**, island records),
- geographical range (**R**, latitudinal extent in mainland Britain),
- distribution (**PS**, proportion of 10 km squares occupied within the range),
- hostplant range (**HT**),
- hostplant abundance (**HA**),
- habitat seral stage (**SS**, seral stage occupied),
- habitat range (**HR**),
- dispersal ability (**D**),
- voltinism (**V**),
- the length of the flight period (**FP**).

Two indices have been computed from these variables. Also an additional variable is included, for comparison with **IR** (island records):

- **EC1**, the sum of all nine variables,
- **EC2**, the sum of the seven strictly ecological variables excluding range and distribution (**R** and **PS**),
- the incidence of species at mainland faunal sources for islands (**FSR**, faunal source records).

Table 8. The frequency of occurrence and ecological attributes of 57 butterfly species on 73 British and Irish islands.

SPECIES		IR	FSR	R	PS	HT	HA	SS	HR	D	V	FP	EC1	EC2
1	*C. palaemon*	0	14	1	1	3	3	3	2	2	1	2	18	16
2	*T. sylvestris*	7	24	3	4	3	3	3	4	3	1	2	26	19
3	*T. lineola*	4	11	2	2	3	4	3	3	3	1	2	23	19
4	*T. acteon*	1	7	1	1	1	3	3	1	1	1	2	14	12
5	*H. comma*	1	5	1	1	1	4	4	1	2	1	2	17	15
6	*O. venata*	10	30	3	3	3	4	3	4	3	1	2	26	20
7	*E. tages*	4	33	4	2	3	4	4	2	2	2	2	25	19
8	*P. malvae*	4	22	2	2	3	4	4	2	2	2	2	23	19
9	*P. machaon*	4	1	2	1	1	1	1	1	2	2	2	13	10
10	*L. sinapis*	3	20	2	1	3	3	3	2	2	2	2	20	17
11	*G. rhamni*	20	37	3	3	2	3	2	2	4	1	4	24	18
12	*P. brassicae*	73	73	4	4	3	4	4	1	4	3	3	30	22
13	*P. rapae*	73	73	4	4	3	4	4	2	4	3	3	31	23
14	*P. napi*	67	73	4	4	3	4	4	3	4	3	3	32	24
15	*A. cardamines*	28	57	4	3	3	4	3	3	3	2	3	28	21
16	*C. rubi*	28	61	4	2	4	3	3	3	2	1	2	24	18
17	*T. betulae*	1	17	2	1	1	4	2	2	2	1	2	17	14
18	*Q. quercus*	9	48	3	2	2	3	1	1	3	1	2	18	13
19	*S. w-album*	4	20	3	2	2	2	1	1	2	1	2	16	11
20	*S. pruni*	0	0	1	1	1	4	2	1	1	1	1	13	11
21	*L. phlaeas*	47	73	4	3	2	4	4	4	3	4	3	31	24
22	*L. dispar*	1	0	2	1	1	1	1	1	2	1	1	11	8
23	*C. minimus*	7	41	4	1	1	2	4	2	2	2	2	20	15
24	*P. argus*	8	23	2	1	4	4	4	2	1	1	3	22	19
25	*A. agestis*	9	25	2	2	4	4	4	3	1	2	2	24	20
26	*A. artaxerxes*	1	7	3	1	3	3	4	1	1	1	2	19	15
27	*P. icarus*	66	73	4	4	3	4	4	4	3	4	3	33	25
28	*L. coridon*	1	6	2	1	1	1	4	1	2	1	2	15	12
29	*L. bellargus*	1	5	1	1	1	1	4	1	1	2	2	14	12
30	*C. argiolus*	21	37	3	3	3	3	3	2	3	4	2	26	20
31	*M. arion*	0	8	1	1	1	2	4	1	1	1	1	13	11
32	*H. lucina*	1	10	3	1	2	2	3	2	2	1	2	18	14
33	*L. camilla*	4	12	2	2	1	3	3	1	3	1	2	18	14
34	*A. iris*	2	12	1	1	2	3	1	1	2	1	2	14	12
35	*A. urticae*	72	73	4	4	1	4	3	4	4	3	3	30	22
36	*N. polychloros*	8	20	2	1	3.5	2	1	1	4	1	2	17.5	14.5
37	*I. io*	64	67	4	3	1	4	3	3	4	2	3	27	20
38	*P. c-album*	15	29	3	4	3	4	3	2	4	3	3	29	22
39	*B. selene*	19	62	4	2	2	3	4	3	2	2	2	24	18
40	*B. euphrosyne*	11	48	4	1	2	3	4	2	2	2	2	22	17
41	*A. adippe*	2	15	3	1	2	3	4	2	3	1	2	21	17
42	*A. aglaja*	44	68	4	2	2	3	4	3	3	1	3	25	19
43	*A. paphia*	9	36	2	2	2	3	2	2	3	1	2	19	15
44	*E. aurinia*	12	52	4	1	2	3	3	2	2	1	2	20	15
45	*M. cinxia*	6	2	1	1	1	3	4	1	2	2	2	17	15
46	*M. athalia*	0	2	1	1	2	3	3	2	2	1	2	17	15
47	*P. aegeria*	35	65	4	3	3	3	2	2	3	4	3	27	20
48	*L. megera*	35	47	3	4	3	4	4	4	3	4	2	31	24
49	*E. epiphron*	0	2	2	1	1	3	1	1	2	1	1	13	10
50	*E. aethiops*	9	30	2	1	3	3	3	2	2	1	2	19	16

Table 8. cont.

SPECIES	IR	FSR	R	PS	HT	HA	SS	HR	D	V	FP	EC1	EC2
51 *M. galathea*	4	20	3	2	3	4	3	3	2	1	2	23	1852
H. semele	48	72	4	1	3	4	4	2	3	1	3	25	20
53 *P. tithonus*	20	33	3	4	3	4	3	4	3	1	2	27	20
54 *M. jurtina*	69	73	4	4	3	4	3	4	3	1	4	30	22
55 *A. hyperantus*	24	57	4	3	3	4	3	3	2	1	2	25	18
56 *C. pamphilus*	53	73	4	4	3	4	4	4	3	4	3	33	25
57 *C. tullia*	24	50	3	3	3	3	3	1	2	1	2	21	15

IR: incidence on 73 islands.
FSR: faunal source records. The frequency of occurrence for species at the mainland faunal sources for 73 islands.
R: number of latitudinal regions occupied within British mainland (scale 1–4); regions, >56˚N, 54–56˚N, 52–54˚N, <52˚N.
PS: proportion of 10 km. squares occupied within range; 1, <25%; 2, <50%; 3, <75%; 4, <100%.
HT: hostplant type; 1, monophagous; 2, oligophagous – 1 species per habitat; 3, oligophagous – >1 species per habitat; 4, polyphagous.
HA: hostplant abundance; 1, substrate dependent; 2, patchy within habitats; 3, ubiquitous within habitat types; 4, ubiquitous and cosmopolitan.
SS: habitat seral stage;1, climax woodland or plagioclimax bog; 2, pre-climax woodland; 3, shrubs, forbs and tall grasses; 4, bare ground, short forbs and grasses.
HR: range of semi-natural habitats occupied 1, <5; 2, <9; 3, <14; 4, <18 (maximum number).
D: dispersal ability; 1, 'closed' populations with little evidence of movement outside colonies; 2, colonial species with evidence of dispersal; 3, 'open' population structure with evidence of frequent movements between habitat units; 4; migrants or vagrants with evidence of long-distance movements.
V: voltinism; 1, biennial or univoltine; 2, univoltine with occasional partial second broods; 3, bivoltine; 4, multivoltine.
FP: flight period length of longest brood; 1, <1 month; 2, 1–2 months; 3, 2–3 months; 4, >3 months.
EC1: ecological index incorporating distribution data = R+PS+HT+HA+SS+HR+D+V+FP.
EC2: ecological index without distribution data = HT+HA+SS+HR+D+V+FP.

Note: *H. morpheus*, *C. croceus*, *V. atalanta* and *C. cardui* excluded.

In a previous work (Dennis & Shreeve, 1991), the ecological variables were scored to generate an index of vulnerability to environmental perturbations; each is described by four ranks such that the lowest score (1) indicates the greatest vulnerability and the highest score (4) the least vulnerability to any environmental change. Only scores for dispersal have been modified to incorporate new data. These scores can also be visualized as direct measures of ecological potential for migration (*sensu* Baker, 1978) and colonization and inverse measures of susceptibility to extinction (see Table 8). For instance, butterfly species with large geographical ranges, denser distributions, more broods and longer flight periods will have a higher probability of migrating to offshore islands than those species with smaller ranges, sparser distributions, single broods and restricted flight periods. Similarly, those species that use a wide range of hostplants and habitats and which have more abundant hostplants are more likely to colonize islands and, once there, have a higher year on year probability of surviving than species that do not. Furthermore, as islands generally have disturbed habitats, owing to high wind speeds, salt spray and active mass wasting along cliffs, species occupying early seral stages should have a higher probability of colonizing and surviving on islands than those requiring resources of later seres. In a number of respects, these variables describe a gradient along a scale of generalist v. specialist strategies, successfully exploited in studies of the differentiation of butterfly communities (Kitihara & Fujii, 1994).

The correlations of species' incidence (**IR**, island records) with species' ranges (**R**) and distributions (**PS**) are close and highly significant (r_s = 0.80 and 0.78 respectively; $P < 0.001$; Table 9). This link between species' incidence on islands and the geography of species on nearby mainlands is confirmed by the very high correlation with faunal source records (**IR** with **FSR**: $r = 0.88$; $P < 0.001$). Butterfly ranges (**R**) and distributions (**PS**) also correlate closely with the ecological index based on seven variables (**EC2**: r_s = 0.65 and 0.79 respectively; $P < 0.001$); these correlations accord with the expectation that ranges and distributions of species reflect their potential for migration, colonization and survival on islands. A more important finding is that the incidence of species on islands (**IR**) correlates highly and significantly with the same ecological index **EC2**: $r_s = 0.81$; $P < 0.001$). If range and distribution are added to this ecological

Table 9. Correlations (Spearman r_s) between the incidence of butterfly species on 73 British and Irish islands and geographical and ecological variables.

	IR	R	PS	HT	HA	SS	HR	D	V	FP	EC1
R	0.804***										
PS	0.776***	0.601***									
HT	0.429***	0.314*	0.421**								
HA	0.482***	0.340**	0.558***	0.410**							
SS	0.223 NS	0.259 NS	0.090 NS	0.150 NS	0.276*						
HR	0.593***	0.575***	0.634***	0.445***	0.576***	0.236 NS					
D	0.690***	0.516***	0.694***	0.215 NS	0.337*	–0.074 NS	0.416**				
V	0.523***	0.423***	0.485***	0.141 NS	0.293*	0.397**	0.308*	0.339**			
FP	0.763***	0.603***	0.584***	0.319*	0.466***	0.255 NS	0.462***	0.634***	0.439***		
EC1	0.866***	0.773***	0.855***	0.574***	0.670***	0.377**	0.791***	0.672***	0.587***	0.740***	
EC2	0.806***	0.654***	0.785***	0.604***	0.732***	0.443***	0.786***	0.620***	0.606***	0.740***	0.974***

***, $P < 0.001$; **, $P < 0.01$; *, $P < 0.05$; NS, not significant.

IR: incidence on islands;
R: range;
PS: proportion of 10 km squares occupied within range;
HT hostplant type;
HA: hostplant abundance;
SS: habitat seral stage;
HR: range of semi-natural habitats occupied;
D: dispersal ability;
V: voltinism;
FP: flight period length;
EC1: ecological index 1 = R+PS+HT+HA+SS+HR+D+V+FP;
EC2: ecological index 2 = HT+HA+SS+HR+D+V+FP.

index, then the correlation is further increased (**EC1**: $r_s = 0.87$; $P < 0.001$). Some 66% of the variation in incidence of species on islands may be accounted for by the selected ecological variables alone. The additional contribution of range and distribution ($r^2 = 76\%$) would suggest that other underlying factors are not being measured, either ecological (i.e., energy environment), historical or human influences.

Individually, all but one of the variables selected correlates positively and significantly with the incidence of species on islands. The exception is habitat seral stage (**SS**: $r_s = 0.22$), though the regression parameters for normalized data ($r = 0.28$, $F_{(1,55)} = 4.7$; $P = 0.034$) are significant. The relatively low correlation of seral stage with island incidence reflects the substantial variation in island incidence for species occupying early seral stages (i.e., higher ranks). The distributions of a large number of species occupying early seral stages are clearly limited by factors other than the stage of vegetation succession. The geographical attributes correlate most highly and, apart from habitat seral stage, the hostplant variables most weakly (**HT**: $r_s = 0.43$; **HA**: $r_s = 0.48$) of the variable suite, but all potentially contribute to the incidence of species on islands. Again, apart from habitat seral stage, modest to high correlations exist between the two biogeographical variables, range and distribution, and the individual ecological variables, indicative that different aspects of the insects' ecology contribute to biogeographical status. This is corroborated, rather than contested, by the generally weaker correlations among the ecological variables (maximum $r_s = 0.47$), the only exceptions being the correlation between dispersal and flight period (**D** with **FP**: $r_s = 0.63$) and betweeen hostplant abundance and habitat range (**HA** with **HR**: $r_s = 0.58$).

The regression parameters confirm the simple rank correlation between species' incidence on islands (**IR**) and the ecological index (**EC2**), as well as illustrating a strong exponential relationship between the two variables (Fig. 9). The scatter of individual species around the regression line provides some indication of the factors responsible for the incidence of species on British islands. The 13 species most commonly found on islands do not score highly on all ecological attributes. For example, none is polyphagous in Britain and three are either strictly oligophagous or monophagous (i.e., *Lycaena phlaeas*, *Aglais urticae* and *Inachis io*). Nevertheless, they all have ubiquitous and cosmopolitan hostplants and are associated with habitats typical of early seral stages. All have a record of being able to migrate over areas lacking in resources, some over very considerable distances (e.g., *Pieris napi*, *Aglais urticae*). The relationship with migration is an important one. It is confirmed by the high correlation between the incidence of species on islands and the sum of seven binary coded variables describing capacity for movement (**IR** on **MI**: $r_s = 0.76$; $P < 0.001$; Tables 7 and 8). Only two species lie two standard errors beyond the regression line. These are *Carterocephalus palaemon* and *Mellicta athalia* which have substantial negative residuals; neither occupies any of the smaller British islands. The inference is that both are particularly demanding of habitat conditions, not measured in the simple ranking of resources, as is clear from autecological research on both species (Warren, 1987a, b, c; Ravenscroft, 1994a, b). Some other species are indicated to be

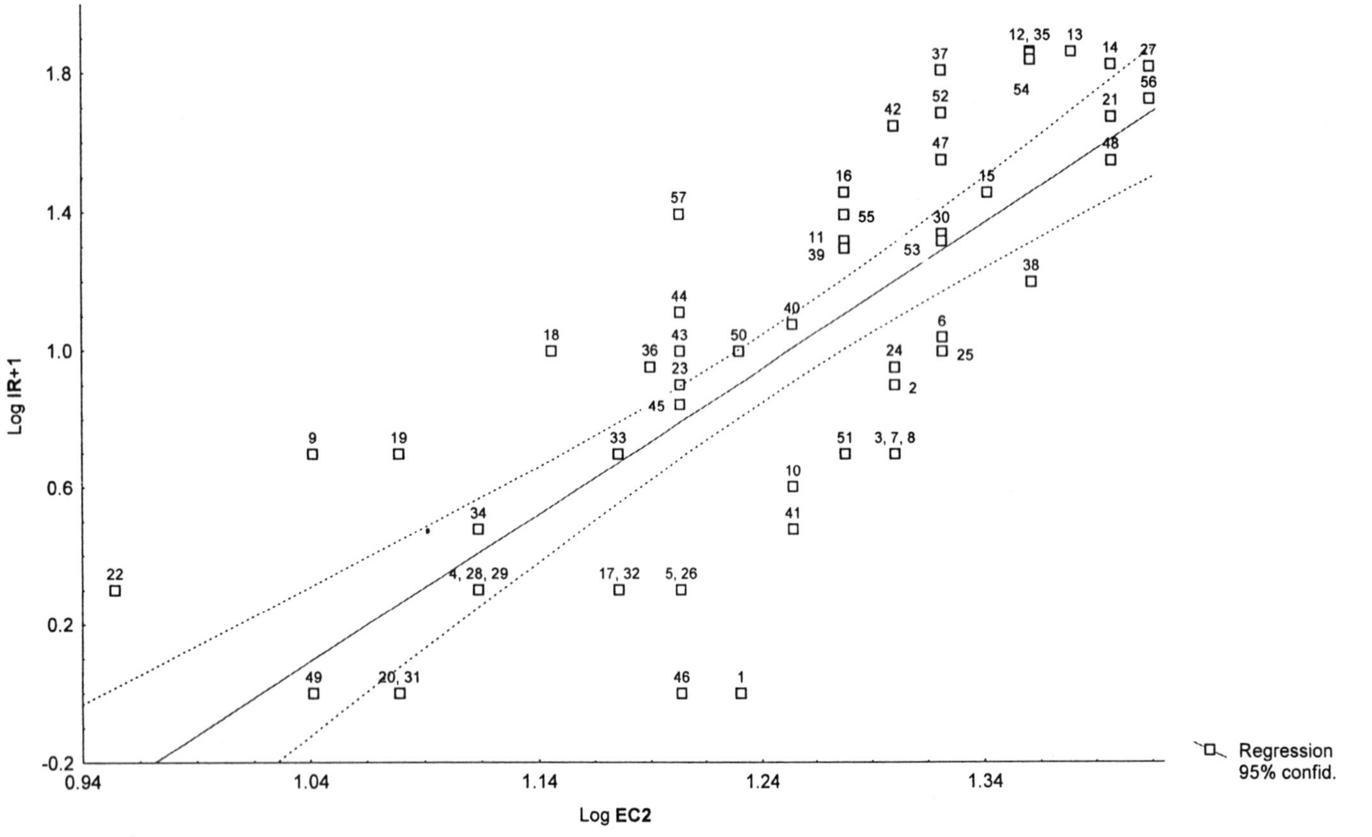

Fig. 9. Relationship of species incidence on islands (**IR**) to an ecological index (**EC2**) based on seven ecological variables. See Table 7 for names of species.

$$\log \mathbf{IR} = -8.95 + 3.98 \log \mathbf{EC2};\ r = 0.77,\ P < 0001.$$

similarly affected (e.g., *Hesperia comma*, *Aricia artaxerxes* and *Argynnis adippe*: Emmet & Heath, 1989). There are also species that have substantial positive residuals. Most are well known for engaging in long distance migrations (e.g., *Aglais urticae*, *Inachis io*). An important exception is that of *Coenonympha tullia* which is found on 24 of the 73 islands analysed (see section I.10).

In previous work it has been shown that the latitudinal distribution of species' richness among butterflies can be mainly 'explained' by summer climate (Turner, Gatehouse & Corey, 1987; Dennis, 1992, 1993). However, this does not explain the distribution of individual species. The present results would suggest that the ranges and distributions of individual species, and their incidence on islands, can be largely accounted for by ecological differences between them. Confirmation of this awaits improved measures of ecological variables on species. The simple ordinal distinctions among species for ecological attributes used in this work by no means fully quantify real differences among them. When this is feasible, it may then be possible to determine the relative importance of different ecological variables on species' distributions.

9. VARIATION OF BUTTERFLIES ON ISLANDS

Different interpretations of geographical variation among British butterflies have greatly influenced evolutionary and historical models for the butterfly fauna of the British islands (cf. Heslop Harrison, 1946d, 1947f, 1948b, 1950c, d; Beirne, 1943a, b, 1947; Ford, 1945; Dennis, 1977, 1992, 1993). As these models are based on very different views as to the biogeography of species on islands around the British coastline, it is important to examine the variation and the models arising from it. Geographical variation in the butterflies of British islands has been described fully elsewhere (Dennis, 1977; Emmet & Heath, 1989). There is insufficient space and no requirement to reproduce such detail here. However, as variation on islands is entered into analyses (see below), a summary of the formally described variation is presented. In the list of islands, subspecies are coded up by superscripted numbers. For some species, unnamed distinct forms and variation have been described; these are also included in the list for islands. No attempt has been made to assess the validity of either the description or the nomenclature of any subspecies; some indication of the acceptability of the various names is provided in Dennis (1977) and Emmet & Heath (1989), but many of the geographical forms await formal revision.

Several issues need to be addressed in any attempt to determine the significance of geographical variation among British butterflies. In some cases, the variation has not even been appropriately designated; for example, precise locations have not been given and types are not readily available for inspection (e.g., *Coenonympha pamphilus scota* Verity). However, the most serious criticisms levelled at naming subspecies of British butterflies are those to do with the process of sampling and quantification. Only rarely are the sampling methods given and any indication that the samples have been taken at random (e.g., *Hipparchia semele clarensis* de Lattin: Howarth, 1971b). Invariably, no attention has been given to variation in space or time; information is often not given on whether the specimens come from a single locality or from several, from a single brood or represent several years. Yet, all this variation has been shown to be extremely large for all so-called subspecies examined closely (Dennis, 1977; Thomson, 1987). Along with the absence of appropriate sampling techniques is the lack of quantification and of any kind of objective comparisons. The variation described has emerged in a haphazard manner; for only a few species have concerted systematic surveys been attempted to describe components of the variation over island groups (e.g., *Maniola jurtina*: Ford, 1964; Dowdeswell, 1981; Brakefield, 1984; Thomson, 1987).

For many of the reasons cited above, the trinomial taxa require revision and the variation, more importantly, detailed survey. There is another feature to the variation. Owing to the way in which detail has accumulated on variation over the years, lack of variation for some species is not evidence for its absence. For example, prior to 1948, Heslop Harrison (1948c) had assumed that *Coenonympha pamphilus* was almost identical everywhere in Britain and had as a result given it little attention. In that year, when in Rhum, he was impressed by the similarity there of variation in *C. pamphilus* to that of *C. tullia*, and named the form *C. pamphilus rhoumensis*. Thereafter, he extended the description of subspecies *rhoumensis* to the Outer Hebridean populations, though specimens of the insect there in 1945 failed to raise any comment from him (Heslop Harrison, 1945c). This extension of descriptions for subspecies from the type-locality to other areas has usually been an even more casual affair than the original description of subspecies, occasionally carried out on the basis of very few specimens. In the list for islands, the uncertainty for descriptions of subspecies is indicated by the placement of codes for subspecies in parentheses.

The absence of a systematic approach to variation in butterfly species on British islands makes interpretation of such variation extremely difficult. The fact that there are more described forms for some species than others (e.g., those for *Maniola jurtina* and *Hipparchia semele* are numerous) is not evidence that these species are more susceptible to the generation of geographical variation than others; they may simply have been subject to more attention. From the manner in which surveys have been carried out there is a dearth of information to test these alternatives. Nevertheless, it is possible to establish several generalizations about the geographical variation described for British islands. Firstly, it affects many species from different families, even those known to migrate regularly over long distances (e.g., *Aglais urticae*: Heslop Harrison, 1937a, 1938a, c; *Pieris brassicae*: Leverton, 1994) as well as those whose movements are more restricted (e.g., *Pararge aegeria*: Heslop Harrison, 1949d). Secondly, when it has been studied, much of the variation has been shown to have a genetic basis (see Dennis, 1993: 109 for references), though none of it would suggest that the differences are large in number or degree. Thirdly, the variation differs enormously in type and direction between species in different families, often in a contradictory manner among species in the same archipelago. This point has been fully addressed elsewhere (Dennis & Shreeve, 1989; Dennis, 1992, 1993). This diversity has been modelled in terms of selection for thermal efficiency, predator escape and mate-location on taxa differing in resting posture, thermoregulation device, activity and habitat affiliation. It was shown that such geographical variation among species increases significantly towards cooler less sunny climates to the north-west of Britain, even on the mainland (Dennis, 1977, 1992). Fourthly, forms are typically described as occupying whole archipelagos (e.g., *Pararge aegeria insula*: Howarth, 1971a; *Maniola jurtina cassiteridum*: Graves, 1930, both in the Isles of Scilly) as well as parts of the adjacent mainlands of Britain and Ireland (e.g., *Pararge aegeria oblita*: Heslop Harrison, 1949d; *Maniola jurtina splendida*: White, 1872; both in north-west Scotland and the Hebrides). Such infraspecific variation, involving a subspecies comprising two or more widely separated but phenotypically identical populations, is described as being polytopic (see maps in Thomson, 1980).

The geography of variation over islands, particularly polytopism, greatly influences the relationships between islands. A previous non-metric ordination of islands (Dennis, 1977: 196) on infraspecies data, when compared to that based on species, revealed several features:

(i) The British mainland 'tree' comprising several regions became extended. Northern Britain pulled away from the south and closer to the northern islands.
(ii) Ireland became significantly more isolated in the plot.
(iii) The Hebridean islands divided into two groups, the northern Highland unit and the southern Strathclyde unit.
(iv) The English islands were drawn southwards into more geographically correct positions.

A similar ordination for 26 islands based on infraspecies data has been carried out for the current file (Fig. 10). On this occasion the ordination of infraspecies data can be compared for different initial configurations: the default Guttman-Lingoes initial configuration and the plot of units in two dimensions from the ordination of species' data. The use of species' data, rather than the default initial configuration, does result in some improvement in the stress values (Kruskal phi, from 12.81% to 12.69%; Guttman's K, from 11.78% to 11.61%) though this is marginal and indicative of small changes involving few units. Two units are affected, the Isle of Man and St Mary's in the Isles of Scilly; Man is drawn closer to Ireland and St Mary's further from Ireland and closer to Skomer and Lundy. The present results compare favourably with the previous plots in 1977:

(i) The British mainland has become extended along dimension 1; the main change is that the northern region (>56°N) has pulled away from southern Britain.
(ii) Ireland has pulled away from the centre of the plot and has drawn the Isle of Man along with it.
(iii) The division among the Hebridean and northern islands is not so dramatic, but some minor clusters are evident (i.e., Arran, Islay, Jura and Colonsay; Mull, Skye and Rhum; Tiree, Coll, South Uist, Lewis-Harris and mainland Orkney) which indicate that distinctions exist among the Hebridean faunas. Shetland is isolated in the plot.
(iv) The southern English, Welsh and Channel islands do not adopt geographically correct positions; their relative placement relates more to the size of the island faunas.
(v) The main impression is of three basic groups – the southern English, Welsh and Channel islands; the Irish islands; and the Scottish islands. All are closely attached to their respective mainland faunal sources.

A very low correlation between affinities among islands on infraspecies' and species' data ($r = -0.03$; not significant) is indicative of how disruptive geographical variation is on the systematic trends in species' data. This is confirmed by the substantial increase in stress values from the two dimensional ordination for the species' data (Kruskal phi, 7.1%; Guttman's K, 7.9%).

Bearing in mind the influence that geographical variation has on relationships among British islands (Fig. 10), a brief examination of the nature of polytopism is

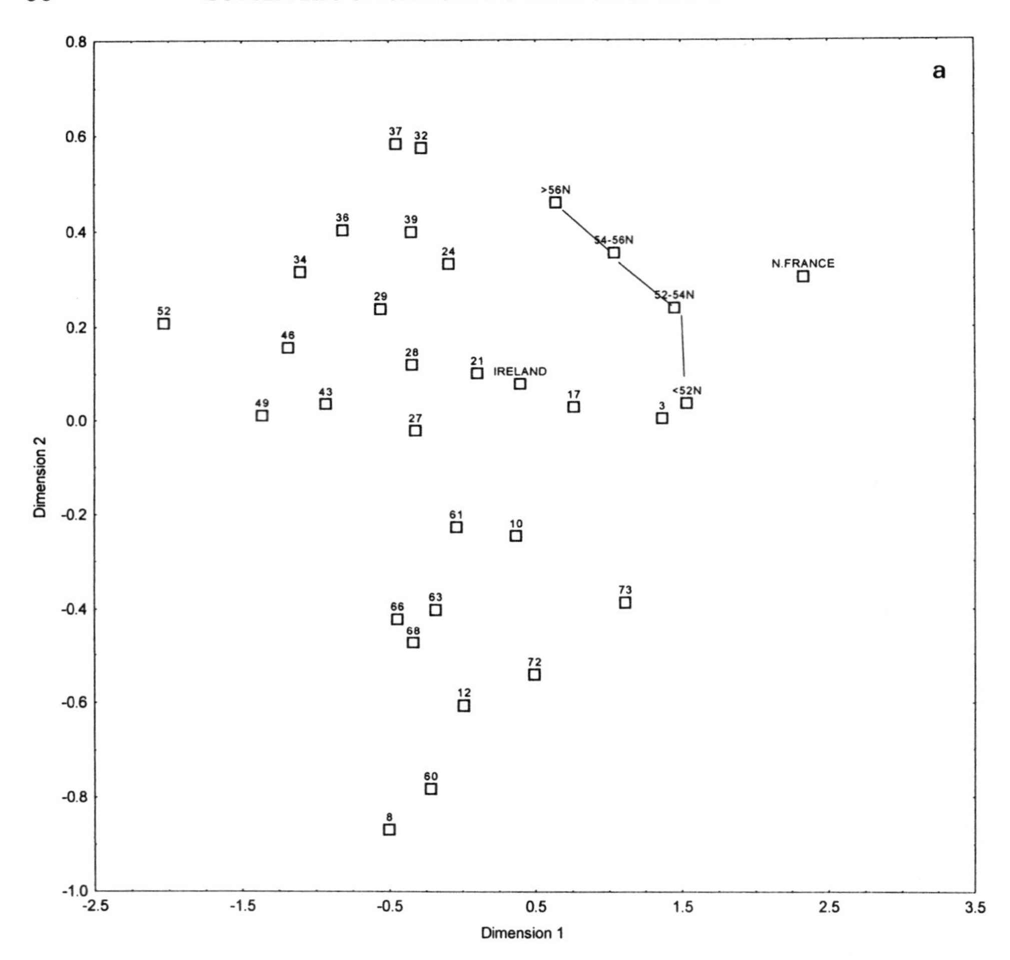

Fig. 10. Non-metric scaling plots (2 dimensions) of British and Irish islands based on (a) species data (Guttman K = 0.079; Kruskal phi = 0.071); (b) infraspecies data (Guttman K = 0.127; Kruskal phi = 0.116). Co-ordinates for islands on species data in

instructive. Subspecies terminology has the effect of ascribing a degree of homogeneity to populations under the same term and distinguishes populations sharing different terms; the distinctions are purported to extend, at least, to placement in space, genetics and evolution. The first point that needs to be made is that there is considerable geographical variation within polytopic subspecies, variation not only between islands, but between populations within islands. Perhaps unexpected, in view of this, is the dearth of endemic subspecies or races decribed for islands. Heslop Harrison (1945c) described a small pale form of *Argynnis aglaja* from Flodday, in the Outer Hebrides, very different in guise from the melanic *A. aglaja scotica* dominating surrounding islands. He also noted two striking populations of *Polyommatus icarus* and *Coenonympha pamphilus* in Allt

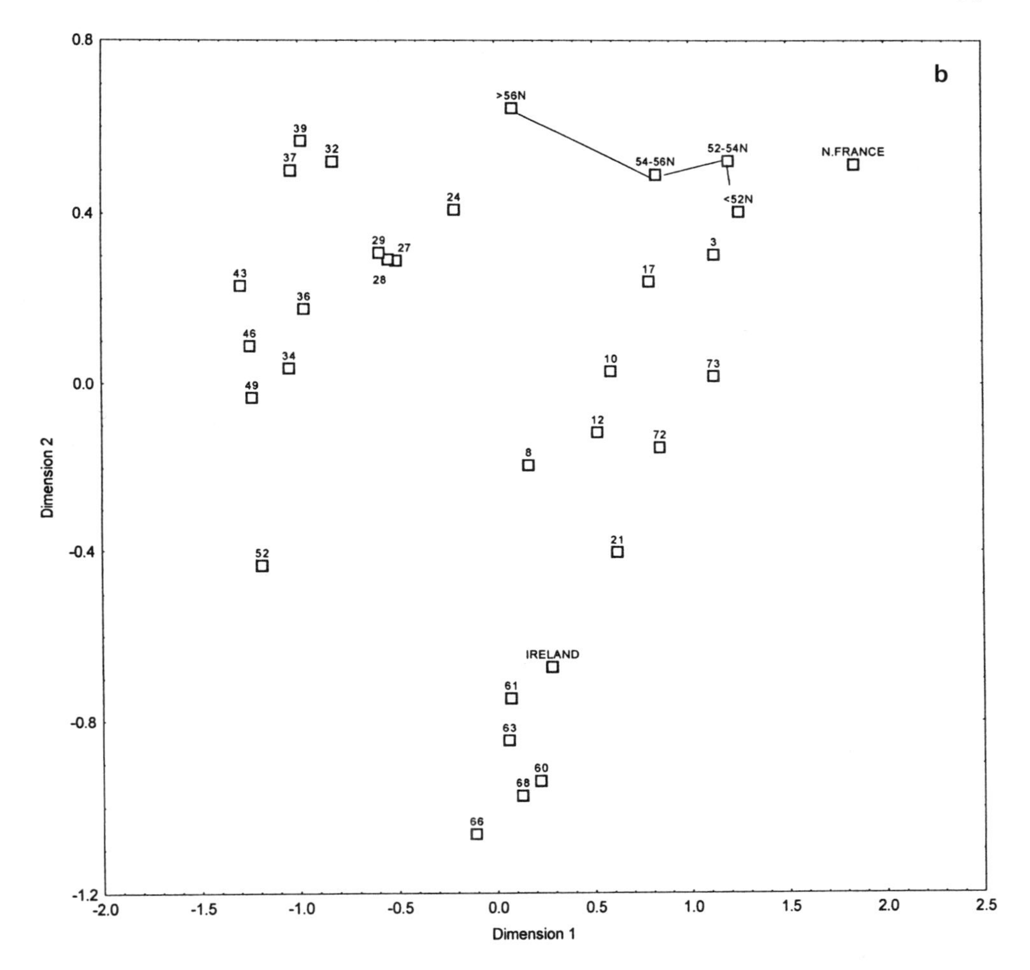

the two dimensional solution provide the initial configuration for infraspecies data. See Table 1 for names of islands.

Volagir gorge on South Uist (Heslop Harrison, 1950b). However, such variation among populations has not generated formal descriptions for populations on single islands. Clearly, it is no greater than variation described for ecological enclaves on the British mainland (e.g., *Hipparchia semele* and *Plebejus argus* on the Great Ormes Head: Dennis, 1977) and in Ireland (e.g., *Erynnis tages baynesi*, Huggins, 1956; *Hipparchia semele clarensis*, de Lattin, 1952). In spite of the absence of island endemic subspecies, a very considerable amount of variation has been disclosed between island populations. To take one example, *Maniola jurtina splendida* has been described as very dull on Pabbay in the Inner Hebrides (Heslop Harrison, 1938a: 19), to be very dark on Taransay (Heslop Harrison, 1938c: 266) and to be particularly large and bright (orange) on Eilean an Tigh in the Shiant islands

(Heslop Harrison, 1953). This considerable variation under a single subspecies term for M. *jurtina* has been confirmed by the detailed surveys of Thomson (1970, 1980, 1987). Such extensive variation is typical of other polytopic subspecies for the Western Isles of Scotland (Dennis, 1977).

Variation among island populations is matched by variation among populations within islands and variation within populations. With respect to this, the designation of subspecies for butterflies in the Western Isles of Scotland has had some unfortunate consequences, since it has led not only to explicit recognition of different subspecies on the same islands (e.g., for *Argynnis aglaja* on Rhum and Raasay: Heslop Harrison & Morton, 1952; Heslop Harrison, 1955b; for *Maniola jurtina* on Scalpay: Heslop Harrison, 1937a) but also incredibly within the same population on one island (e.g., *Coenonympha tullia* on Rhum, Lewis and Harris; Heslop Harrison, 1950c; Heslop Harrison & Morton, 1952). This raises the second point about polytopic variation on British islands, viz., the immense amount of temporal variation. Heslop Harrison was well aware of its existence but failed to quantify it or to interpret it correctly. It led him to describe spatial and temporal changes in wing pattern phenotypes on islands in terms of taxonomic labels (Heslop Harrison, 1947f; for a similar usage see Ford, 1957: 292). To summarize, some general features can be identified for variation among polytopic subspecies in north-western Britain:

(i) Distinct phenotypes have been observed which form the basis of what are probably statistically significant distinctions among groups of populations; this has led to the formal description of subspecies.
(ii) Much variation exists between populations under the same taxon, especially between those on different islands, but also between populations on the same island. Considerable overlap exists between such populations, apparently even between those designated as belonging to different subspecies. There are examples of variation described for northern populations occurring in southern populations (e.g., Cornish, Surrey and Isle of Wight *Eurodryas aurinia*: Hodgson, 1935; Johnson, 1955; Huggins, 1972; Durham and Cornish *Argynnis aglaja*: Heslop Harrison, 1958a; W. G. Tremewan, pers. comm.; Sheffield *Polyommatus icarus*: Fearnehough, 1937).
(iii) There is very considerable temporal variation within populations, within broods as well as between seasons. This variation can transend the description of phenotypes for different subspecies in a number of species (e.g., *Maniola jurtina*, *Argynnis aglaja*, *Coenonympha tullia*). As a result of these observations, it is evident that polytopic clusters for so-called subspecies are extremely variable abstractions.

It was mentioned earlier that the different interpretations of variation presented by island subspecies has greatly influenced explanations for its evolution. Two contrasting models have been suggested. The first interprets each subspecies as originating from a single unit modified in isolation, which has subsequently dispersed and, in turn, has founded populations on different islands; this may be called the glacial refugium model. The alternative model envisages the colonization of islands by founding individuals, the populations of which assume similar phenotypes owing to the commonality of similar selection pressures. For convenience, this may be called the Holocene selection model. At face value, these models do not seem to be greatly different. When summarized as above, the main distinction is that in the former phenetic differentiation defining the infraspecies units occurs prior to their isolation on islands; in the latter it occurs subsequent to isolation. This rather underscores the differences; in fact, much more is involved. The models incorporate very different concepts of time and history, and differ markedly in dynamics, dynamism in phenetic changes and population integration.

The fundamental difference between the two models is the significance attributed to the rank of subspecies. In the glacial refugium model, Heslop Harrison (1946d, 1947f), Beirne (1943a, 1947) and Ford (1945) assumed that subspecies evolved over long periods of time under extreme conditions. As such, the subspecies were considered to have evolved in off-shore refugia during lower sea-levels of the last or preceding glaciations. This model also assumes the need for land connections, and as such this greatly influences the interpretation of butterfly records on British islands. It effectively proposes a fixed or static view of island biogeography. Moreover, the model could not explain the variation in subspecies. This was simply attributed to isolation during glacial phases, and discordant phenotypes on one island (or in one population) were interpreted as the mixing of infraspecies taxa and of great biogeographical significance. This model has been reviewed in detail elsewhere (Dennis, 1977, 1992; Dennis & Shreeve, 1989). It has been rejected by us:

(i) The alleged Pleistocene refugia for resident British butterflies did not exist, or conditions on them – in fact anywhere in the vicinity of the British islands – were too extreme for their survival.
(ii) Land-bridges were not available for many islands, and are not necessary for immigration to and colonization of islands.
(iii) Infraspecies differentiation is not great, but inter- and intra-population variation is substantial. Thus, the interpretation of variation within islands and within populations on islands as a mixing of taxa is invalid; it had no theoretical validity anyway.

The alternative Holocene selection model, based on a more or less parallel response of island populations to similar selection pressures induced by changes in environmental conditions to the north and west of Britain, does not depend on limited colonization events dictated by the exigency of land-bridges. A reconstruction of their history, based on the geography of their resources, shows that butterfly species very probably colonized the British islands in the early Holocene (c.10 ka BP). Nevertheless, there have undoubtedly been frequent extinctions on islands, especially on small ones, and equally frequent recolonizations from the mainland and adjacent islands. This accords with the expectation of island biogeography notions. There are several other important areas of biological theory and fact with which this model tallies. First, gradients in species' richness, phenology, physiology and population dynamics suggest that selection gradients marginalize butterfly species to the north and west of Britain

(Dennis, 1993; Shreeve, Dennis & Pullin, 1996). Second, there is evidence for very rapid changes in phenotypes in butterfly populations, especially when they are the product of limited founder events (e.g., *Plebejus argus* in the Dulas valley, North Wales: Dennis, 1972, 1977). Third, there is growing evidence that many of the attributes distinguishing the so-called subspecies are simple polymorphisms or are polygenic, susceptible to selection and rapid change (see references in Dennis & Shreeve, 1989; Dennis, 1993). Fourth, the variability of populations corresponds with the high degree of plasticity for phenotypes under the extremely variable environments of north-west Britain (Thomson, 1973, 1987); even some southern populations periodically simulate variation found in northern populations (see above). In view of these observations, differentiation among populations can be accounted for by differential gene flow, founder events, drift and disruptive selection, whereas homogeneity can be related to migration and gene flow, and parallel selection regimes.

10. HISTORICAL CONSIDERATIONS

The results of this survey indicate that both species' richness and the incidence of species on islands can be largely 'explained' by current ecological factors though the previous section, on variation among island populations, would suggest that historical influences operating over the Holocene cannot be ignored. Certainly, the high correlations of island incidence with ecology clearly question the significance of historical influences. Prominent specific issues are:

(i) Do the current findings challenge the model for early Holocene entry of species into the islands (Dennis, 1977)?
(ii) Have the processes of migration, colonization and extinction on islands been uniform over time? If not, then what might we expect of the rates and magnitudes of such events?
(iii) Is it possible to identify any historical 'signals' in the current ecological data set?

Before examining aspects of the historical model for entry of species into the British islands relating to the current findings, it is worth elaborating on two statistical points. First, although high levels of variance for species' richness and species' incidence are accounted for by ecological variables, there is much residual variance that has not been allocated (20% for species' richness and 24–36% for species' incidence). Some of this could relate to history. Second, it is not known to what extent ecological variables co-vary with historical influences, for example those influencing the size of faunal sources on the adjacent mainlands or those determining the number of species on archipelagos, a faunal source to which affiliated islands have access and to which they contribute. The point needs emphasis that 'explained' variance in regression models gives maximum estimates of co-variation and does not in fact provide an explanation at all. Biological and ecological explanation still depends on logic and common sense.

The historical model for entry of butterfly species into Britain simply makes use of knowledge on current resources for species (i.e., climatic tolerance, hostplants, habitats) to predict when conditions were first suitable for them to establish themselves in the islands. The application of this method is made necessary by the lack of butterfly fossil records. There is insufficient space to describe the findings in any detail (see Dennis, 1977, 1992, 1993), but some of the significant conclusions can be enumerated:

(i) The last (Devensian) maximum glaciation (circa 20 ^{14}C ka BP) was an effective *tabula rasa* for all current species occupying the islands.
(ii) Part of the current fauna (24%) entered in the Late Glacial and survived in parts of the islands during the severe cold period of the Loch Lomond readvance (circa 11 to 10 ^{14}C ka BP).

(iii) Most species (76%) first arrived during the early Holocene (circa 10 to 9.5 ^{14}C ka BP).
(iv) Changing conditions (i.e., forest growth and its removal, human impact, climatic changes) caused large changes in ranges and distributions of species during the present Holocene interglacial. Such changes have undoubtedly resulted in some extinctions among butterflies in Britain as they have among other organisms (Simmons & Tooley, 1981).
(v) Most phenetic modifications and many other adaptations (e.g., physiology and phenology) among British butterflies have occurred during the Holocene as a response to regional environments; their evolution has been extremely rapid as no doubt is their response to changing conditions (Dennis, 1977, 1993).

The current findings for butterfly species on British islands do not contradict these conclusions, but they do enhance an understanding of processes operating during the Holocene. First, current data on migration and movement (Table 7) support the deduction that land connections are unnecessary for butterflies to have crossed to islands from the Continent and mainland Britain. Second, these results would indicate that migration to, colonization of and extinction on the offshore islands has occurred throughout the Holocene. There is every reason to expect that these processes are partly explained by island biogeography theory, that is, the rates of migration and colonization equate to island isolation and that extinction rates are determined by island area. During the Holocene, it is most likely that large islands act as sources for archipelagos and that whole archipelagos or single islands are serviced from adjacent mainland shores. Changing climatic conditions, especially the summer thermal environment, largely accounts for the number of species at mainland sources. However, ecological parameters determine which species are found at faunal sources.

The question of whether, during the Holocene, processes of migration, colonization and extinction have been uniform is an important one. Recent population models, especially those developed to simulate metapopulations, tend to treat such processes as constant (Hanski, 1994). Yet, observations on natural populations, as well as theoretical modelling of these processes with respect to changing conditions would suggest that they vary in time as well as in space. The reconstruction of colonizations and extinctions during the Late Glacial and Holocene strongly indicates marked variation in rates and magnitude of events. However, recent history provides direct observational evidence from three different sources:

(i) migration;
(ii) colonizations and range expansions;
(iii) population extinctions and range contractions.

All three have have been shown to be periodic and to differ in magnitude with time for different species. There have been seasons when migration rates have been unusually high for so-called sedentary species with closed population structures, such as 1976 (e.g., *Quercusia quercus*, *Cupido minimus* and *Eurodryas aurinia*: Horton, 1977; Holloway, 1980). Waves of colonization, also involving differential migration rates for species, have led to sudden range expansions such as those which took place during the period from 1988 to 1992 (e.g., *Thymelicus sylvestris*, *Pararge aegeria*, *Pyronia tithonus* and *Celastrina argiolus*) (Hardy, Hind & Dennis, 1993). Similar but more extensive episodes of range expansion have been recorded for *Ladoga camilla* and *Polygonia c-album* in the 1930s and 1940s (Pollard, 1979; Pratt, 1986–87). Range contractions have also been demonstrated to be episodic, the result of increased population extinctions. Colonies at the edge of the range for species are particularly prone to this process (Birkett, 1995). In Durham and Northumberland, episodic range contractions as well as expansions have affected a number of species from the 1850s to the present day (e.g., *Lasiommata megera*, *Argynnis aglaja* and *Anthocharis cardamines*: Heslop Harrison, 1954b, 1958a; Dunn & Parrack, 1986).

Evidence is accumulating that underlying the episodes of increased rates in migration, colonization and extinction are climatic events (Dennis, 1977; Hardy *et al.*, 1993). Data from the butterfly monitoring scheme tend to confirm these observations (Pollard & Yates, 1992, 1993; Pollard, Moss & Yates, 1995). The inference is that migration, colonization and extinction of species on islands will tend to be episodic and to be climate- and weather-driven. This would certainly follow from the relationship between size of faunal sources and the summer heat environment. What is most unfortunate is the lack of quantitative data on individual species for rates and magnitude of annual migrations and range changes. Such data are urgently required if these processes are to be effectively modelled. It is perhaps not unreasonable to assume that population processes leading to gains or losses of colonies behave similarly to other natural events that are subject to the vagaries of climate and weather, for instance droughts and floods. In this way, migrations, colonizations and extinctions may be treated as random events, the underlying premise being that such events, during any period, constitute a sample from an indefinitely large 'population' (series) in time. Thus, if in 30 years of records, the largest annual migration event recorded for a species was of a certain size, it is probable that the next 30 years will also contain a migration of equal magnitude. The recurrence interval (I) for events described by a single value each year is given by:

$$I = N + 1/M$$

where N is the number of years and M is the rank of the individual item in the array.

When the annual maximum migration event and probability are plotted on logarithmic probability paper, an approximately linear relationship is described. Extending the line beyond the records available allows the prediction of more extreme events. Alternatively, annual maximum migration events may be plotted against the recurrence intervals on semi-logarithmic co-ordinates (Gumbel paper), again producing a quasi-linear relationship (Fig. 11):

$$D = a + b \log I$$

where D is the maximum migration event in any year and a and b are regression coefficients.

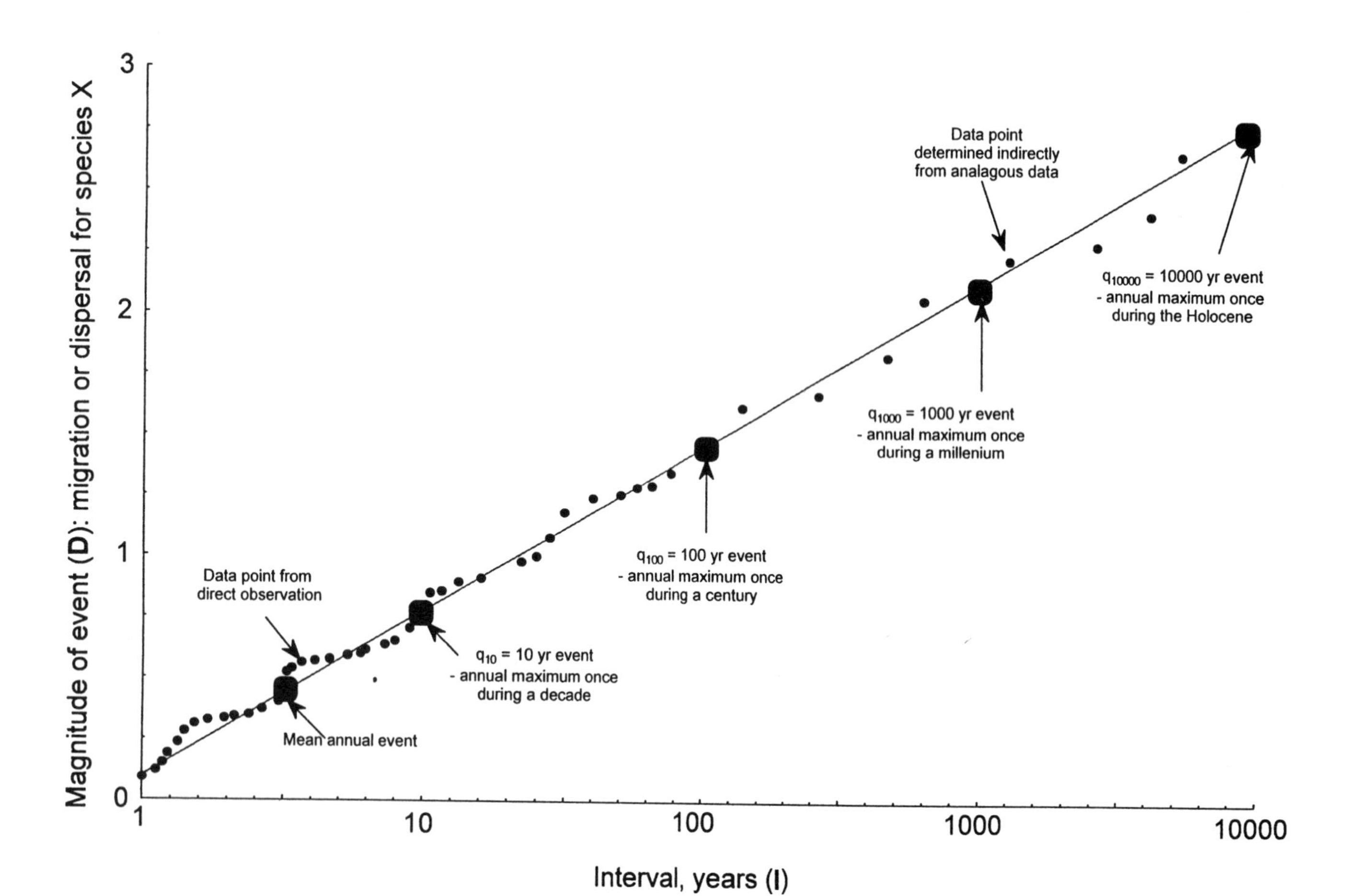

Fig. 11. A graphical model relating the size of annual migration events (D; e.g., range of movements or mass of vagrants per unit distance) to their recurrence interval (I, return period in years). See text for explanation.

The exact form of the relationship is unknown for butterfly populations, but one inference supported by qualitative data is that higher magnitude events of migration, colonization and extinction do occur and much less frequently than lower order ones. It is not surprising that high magnitude events are rarely observed by field ecologists during typically short-term research programmes, who then tend to model such processes as colonization and extinction as being uniform in time. Yet, these high magnitude events may have a substantial impact on range changes and island occupancy, whereas many years of low magnitude events may have no significant effect at all. A corollorary of this comes from hydrology where erosion, associated with river discharge, can be negligible during many years of low discharge but devastatingly large during a single high-magnitude flood. An example worth quoting, as it has significantly altered the landscape of one of Britain's most important butterfly sites, is the erosion of the Great Ormes Head in North Wales caused by a single days rainfall during 10 June 1993; some 137 mm of rain fell in two hours causing the removal of more than 1,900 m^3 debris from 322 ha, an event likely to occur less than once in over a thousand years. It resulted in deep ravines and gullies, massive cliff falls and mass wasting of tons of exposed sediment. This storm may have substantially increased the mortality among butterfly populations on the headland, particularly those of the two races of *Hipparchia semele* and *Plebejus argus* which emerge during mid June; transect data indicate that their populations may have dropped by half and a third respectively in 1993 (Dennis & Bardell, 1996). The hot dry summer of 1995 provided another extreme event though not of the same magnitude. During this summer both species were found as much as 1 km away from known colonies and in cooler habitats than usual (Dennis & Bardell, 1996).

If butterfly migration events take on the form of the inverse relationship modelled between magnitude and recurrence interval, then there is one important consequence for island records. Assuming that recorders are there to make observations, the probability of a species being recorded on an island increases with time. It is important to understand that the records for British islands have been made over at least 50 years. Given time, one may expect that all the species occurring at the nearest mainland faunal source will eventually be observed on an island. This will occur earlier for islands close to shore (e.g., Hilbre; see Appendix 2) than those distant from it. The fact that none of the grass-feeding skippers has yet been recorded from Ireland, nor other species that may be expected to occur there (e.g., *Boloria selene*), indicates that there are measureable limits to migration and colonization for particular species operating over the longer term.

Although many factors may have contibuted to species' richness and species' incidence on Britain's offshore islands during the Holocene, it is another matter to detect historical 'signals' in biogeographical data. Later events in time may substantially obscure earlier patterns and there are pitfalls to the interpretation of residual elements in distributions. Determining the historical significance for distributional data is greatly frustrated by the lack of a fossil record for the Lepidoptera. Nevertheless, historical 'signals' may be identified from two sources:

(i) as residuals from regressions of incidence data on ecological attributes;
(ii) from the comparative genetics of island and mainland populations.

In the first case, it is important to appreciate the limitations of using residuals to detect species that fail to match ecological expectations. Not only may there be critical attributes unaccounted for but, for many species, current distributions which fit ecological predictions may be largely in harmony with historical patterns and processes. As discussed above, correlation of species' incidence with ecological variables does not preclude long-term survival of populations on archipelagos dating to the early Holocene. All but one British species belong to what has been termed the widespread European 'extent' faunal element and have a wide tolerance of climatic conditions (Dennis, Williams & Shreeve, 1991); islands also provide particularly suitable habitats for them. An exceptional species is *Coenonympha tullia*. This butterfly occupies many more islands than would be expected from its ecology. It is generally described as being oligophagous and it is a hostplant-habitat specialist on *Eriophorum-Erica* peat mosses. Besides being restricted in habitat use, it is limited in voltinism and the length of the flight period. Moreover, it is not known to engage in extensive ex-habitat movements, though this may be underscored (Eales, 1995). As such, the factors resulting in its presence on islands may differ from those of other species with similar incidence levels (e.g., *Anthocharis cardamines*, *Aphantopus hyperantus*). The habitat of *C. tullia* forms part of the oligotrophic blanket and raised bog community. This represents a self-perpetuating climax development under the cool wet climates characteristic of north and west Britain dating to the early Holocene (<7.5 ka BP: Walker, 1970; Barber, 1993).

Consequently, it is feasible that populations of *C. tullia* have been maintained on northern archipelagos since their colonization in the wake of Late Glacial ice recession (circa 10 ka BP: Dennis 1993).

Populations of butterfly species on some island archipelagos, typically the Hebrides and Orkneys, have been inferred to have survived there or nearby for long periods of time from the pattern of differentiation in phenotypes, voltinism and ecology (e.g., *Argynnis aglaja scotica* and *Eurodryas aurinia scotica*: Heslop Harrison, 1946d, e, 1947f, 1950c; Beirne, 1947). There are dangers in assuming specific periods of antiquity for populations from phenetic differences; such variation was at one time attributed glacial relict status from British refugia. Not only has this line of reasoning been discredited, but it is highly probable that such variation can evolve very rapidly and independently on different islands in response to similar selection regimes (Dennis, 1977, 1993; Shreeve, Dennis & Pullin, 1996). Nevertheless, such variation as exists on different island groups presents an opportunity for research into the comparative genetics of different species. In this respect, the profound environmental changes during the Devensian glaciation and Late Glacial in Britain at least provides base lines (18 and 10 ka BP) for calibrating DNA variation between the different populations, which can be matched against quantitative appraisals of infraspecific variation on the islands.

11. SUGGESTIONS FOR FUTURE WORK

A main objective of this publication is to encourage work on islands that will throw light on the ability of butterflies to migrate across resource-vacant barriers such as open sea. It is also hoped that this work will encourage more intensive research into the colonization and extinction processes of butterflies, moths and other insects (see Thomas, 1994, 1995) and lead to the accumulation of data over longer periods for the construction of more realistic colonization and extinction probability distributions. Where there is the opportunity for systematic recording on islands, a great deal of useful information can accumulate by applying simple methods, such as the transect technique used by the ITE Butterfly Monitoring Scheme (Pollard & Yates, 1993). This can be enhanced using other sampling techniques for early stages. The issues regarding data collection are discussed in Appendix 2, the problems illustrated by records from Hilbre Island off the Wirral peninsula in Cheshire. Island archipelagos offer rather special opportunities for research into colonization and extinction, as for instance in the case of *Melitaea cinxia* in the Åland islands off Finland (Hanski, Kuusaari & Nieminen, 1994); they present at least two distinct levels of isolation, of habitats within islands and of individual islands from each other and adjacent mainlands. There is, of course, also the opportunity of investigating the impact of the structure of habitats, that is of varying resource geography *within* habitats, on movements.

The present work suggests that movement in butterflies is greater than previously considered. This may be largely a matter of scale. Data for the current survey have necessarily accumulated over years, most over the last three decades, whereas conclusions about movements between terrestrial habitats mainly derive from autecological surveys of less than three years' duration. Therefore, there is a greater probability that the current work has included more and larger high-magnitude events. However, there is another possibility and that is that migration over water surfaces is fundamentally different from that over land. This is a question that deserves careful investigation. Land surfaces, even though they may lack hostplants and fail to provide suitable habitats for colonization, may nevertheless provide opportunities for resting, nectaring and roosting, and encourage host-searching and mate-searching behaviour. Such intervening opportunities for vagrants between a source and potential destination could enhance long-distance migration by increasing longevity. But, they may instead retard migration and deflect it from a linear path. On the other hand, extensive water surfaces present no such inducements to deviate or to be delayed. In other words, suboptimal habitats (i.e., a subset of resources), instead of constituting effective stepping stones which could increase the probability of colonization in a vacant habitat, may isolate the vacant habitat further by reducing the life-time distances that vagrants travel. These relationships are easily described by orthodox gravity model equations and are capable of being tested in the field.

In Britain there is also much research that needs to be done on the ecology and adaptability of species, especially in northern and western islands. Much of this work falls under the heading of marginality, i.e. how populations of species adapt to conditions at the margin of their geographical ranges. A full discussion of the issues is provided in Shreeve, Dennis & Pullin (1996). The threat of rapid climatic change for butterfly populations and their adaptability, owing to enhanced greenhouse conditions, substantially increases the urgency as well as the potential for such research (Dennis, 1993). Much basic autecology is also needed on butterfly populations in the northern islands. Many species have been thoroughly studied in southern Britain; but, for western and northern areas still little has been added to our knowledge about butterfly ecology, even basic details of hostplant choice and habitat structure, since Heslop Harrison's work in the 1930s to early 1950s. An outstanding problem is the need to make sense of wing pattern and other variation found in northern and western populations, and of the underlying genetic mechanisms. Although models have been drawn up to explain this variation in terms of contemporary agents (Dennis & Shreeve, 1989; Dennis, 1993), they have not been tested in the field, and yet they were developed with this purpose in mind.

12. CONCLUSIONS AND SUMMARY

1. Results are presented of analyses on butterfly records for 73 British and Irish offshore islands. The islands have been selected on the basis of records for four migrant species: *Pieris brassicae*, *P. rapae*, *Vanessa atalanta* and *Cynthia cardui*.
2. Species' richness on islands is accounted for mainly by the size of the faunal source of nearby islands with larger faunas ($r^2 = 0.55$) and by the nearest faunal sources at mainland locations ($r^2 = 0.45$). For the present data set, island area (max $r^2 = 0.14$) and island isolation (max $r^2 = 0.10$) are far less important variables in accounting for the number of butterfly species on islands. The size of these correlations for isolation and area, however, may be affected by the cumulative nature of records for butterfly species on islands and the fact that they are based purely on observations and not restricted to breeding records. In a short-term synoptic survey of breeding records, small islands would be more clearly distinguished from large islands. The size of the faunal sources on nearby mainlands and the latitude of sources are closely correlated ($r^2 = 0.86$); their relationship to summer climatic conditions is well known.
3. Modelling affinities among British islands by so-called bootstrapping procedures (using samples drawn from the data to assess the influence of faunal richness and geography) confirm a significant geographical influence on island faunas. Butterfly faunas in the British islands are distinctly nested at different levels. In turn, faunas on small islands, large islands, local mainland sources, regional sources and northern faunas are subsets of larger adjacent units and ultimately nested subsets of a regional fauna in southern Britain and France. This emphasizes the strong influence that geography has on butterfly faunas throughout the British islands. However, exceptions do exist; *Erebia aethiops* and *Coenonympha tullia* on islands and *Erebia epiphron* and *Aricia artaxerxes* on the British mainland do not contribute to southern faunas. It also demonstrates that local faunal sources (for islands) on the British and Irish mainlands differ little from the fauna of the much larger surrounding regions.
4. Data on butterfly movements demonstrate that much ex-habitat migration occurs, indicating the capacity of the majority if not all the British butterfly species to migrate to nearby offshore islands. Evidence comes from at-sea records, island vagrants, ex-habitat vagrants, suburban garden records, city central-business-district records, long-distance movements and mass movements and expansions in ranges. There are clear indications in the literature that both the volume and magnitude of migration for individual species vary periodically, much as other natural phenomena. If so, then it may be possible that they co-vary with extreme weather events (e.g., droughts) and can be modelled in the same way. Another inference is that cumulative records

of species for islands, especially near-shore islands, would with time increasingly match the list of species at the nearest mainland faunal sources.

5. Data on movements suggest that describing the population structure of species as being 'open' or 'closed' is inappropriate, especially as it implies a fixed status for species in space and time based on their mobility, for which suitable data are lacking. An alternative would be to apply terms for types of metapopulations (Harrison, 1991), as clearly many species belong to more than one metapopulation type in different locations and at different times.
6. The incidence of species on islands correlates very closely with the ranges and distributions of species on the British mainland and with their incidence at mainland faunal sources. All these geographical variables also correlate very closely with an ecological index ($r^2 = 0.66$) which purports to describe the capacity of species to migrate to and colonize islands and to resist extinction. There is also a very high correlation between the incidence of species on islands and a movement index ($r^2 = 0.58$). The latter is based on data from seven different sources (see point 4 above). It is suggested that differences in ecology between the species largely account for their incidence on islands as well as their geographical ranges and distributions.
7. Relationships among island faunas are affected by infraspecific variation on the islands, leading to clusters of islands comprising populations which share wing pattern variation. However, much of this variation is not great in terms of genetic differentiation and can be modelled on contemporary environments and selection regimes. Independent evidence rejects long term isolation in glacial refugia and supports evolution in recent Holocene environments.
8. Historical components (evidence of occupation throughout the Holocene) in island species' incidence are not prominent in the data. There are indications of historical signals in the residuals from the regression of island species' incidence on species' ecology, for instance for *Coenonympha tullia* which occurs on more islands than expected and which occupies a stable habitat that has persisted on northern and western islands throughout the Holocene. Phenetic variation in butterfly populations also contains evidence of historical influences (see Dennis, 1993). An important artefact of the existence of modern day spatial patterns and processes amid steep environmental gradients is their potential for mimicking historical patterns and processes; as such, history could be substantially underscored. Much of butterfly species' geography in Britain is determined by summer climate working on ecological differences among species. Although absolute values have changed throughout the Holocene, regional contrasts in climate have remained much the same. Species' incidence on islands seems to relate as much to regional as to local faunal species' pools on mainlands (point 3 above). Numerous elements of such regional faunas, as well as those of large islands, have probably persisted throughout much of the Holocene; these intermittently can have served each other and nearby smaller islands. As islands provide highly suitable habitats for the majority of British species which require open (early seral) conditions, the potential for long-term persistence of species on islands cannot be ignored.

II. RECORDS OF BUTTERFLIES FROM BRITISH AND IRISH OFFSHORE ISLANDS

1. CHECK LIST OF BRITISH AND IRISH BUTTERFLIES

The scientific names and vernacular names of species in the check list follow Emmet & Heath (1989) and Dennis (1992), although the former may not be strictly valid (cf., Kudrna, 1986; Hesselbarth, Van Oorschot & Wagener, 1995; Nässig, 1995). The designation of higher taxa also largely follows Emmet & Heath (1989), though with the inclusion of an additional subfamily in the Hesperiidae. Much disagreement surrounds the notation of higher taxa in the Hesperioidea and Papilionoidea; this remains largely unresolved (cf. Ackery, 1984; Scott, 1985; Brock, 1990; Scoble, 1992). Butterflies occurring as native species, residents formerly native but now extinct and common and infrequent immigrants are included in the list as are those which may have been accidentally introduced but which now breed in the islands (e.g., *Heteropterus morpheus*). Exceptionally rare immigrants, rare accidental introductions and adventives into the British islands are excluded from it (see Appendix 1 and classification in Emmet & Heath, 1989).

Abbreviations and notes:

R resident species;
E residents now known to be extinct;
M frequent immigrant;
I infrequent immigrant;
O immigrants which may overwinter;
B immigrants which may breed;
D deliberate introduction;
A accidental introduction;
? status (identification or native record) uncertain; ? following another symbol indicates uncertain data;
BM British mainland (including islands off England, Wales and Scotland) and Channel islands;
Ire Ireland and its islands (Baynes, 1964; Emmet & Heath, 1989);
+ species not determined, and could be *Pontia edusa* (Fabricius, 1777) (see Wagener, 1988).

Superfamily HESPERIOIDEA LATREILLE, 1809

Family Hesperiidae LATREILLE, 1809

Subfamily Heteropterinae AURIVILLIUS, 1925

Carterocephalus palaemon (PALLAS, 1771) Chequered Skipper (BM: R; E & D in England)
Heteropterus morpheus (PALLAS, 1771) Large Chequered Skipper (BM: A in Jersey only)

Subfamily Hesperiinae LATREILLE, 1809

Thymelicus sylvestris (PODA, 1761) Small Skipper (BM: R)
Thymelicus lineola (OCHSENHEIMER, 1806) Essex Skipper (BM: R)
Thymelicus acteon (ROTTEMBURG, 1775) Lulworth Skipper (BM: R)
Hesperia comma (LINNAEUS, 1758) Silver-spotted Skipper (BM: R)
Ochlodes venata (BREMER & GREY, 1853) Large Skipper
subsp. *faunas* (TURATI, 1905) (BM: R)

Subfamily Pyrginae BURMEISTER, 1878

Erynnis tages (LINNAEUS, 1758) Dingy Skipper
subsp. *tages* (LINNAEUS, 1758)[1] (BM & Ire: R)
subsp. *baynesi* HUGGINS, 1956[2] (Ire: R)
Pyrgus malvae (LINNAEUS, 1758) Grizzled Skipper (BM: R)

Superfamily PAPILIONOIDEA LATREILLE, [1802]

Family Papilionidae LATREILLE, [1802]

Subfamily Papilioninae LATREILLE, [1802]

Papilio machaon LINNAEUS, 1758 The Swallowtail
subsp. *britannicus* SEITZ, 1907[1] (BM: R)
subsp. *gorganus* FRUHSTORFER, 1922[2] (BM: I, O, B)
Iphiclides podalirius (LINNAEUS, 1758) Scarce Swallowtail (BM: I, B)

Family Pieridae DUPONCHEL, [1835]

Subfamily Dismorphiinae SCHATZ, [1886]

Leptidea sinapis (LINNAEUS, 1758) Wood White
subsp. *sinapis* (LINNAEUS, 1758)[1] (BM: R)
subsp. *juvernica* WILLIAMS, 1946[2] (Ire: R)

Subfamily Coliadinae SWAINSON, 1827

Colias hyale (LINNAEUS, 1758) Pale Clouded Yellow (BM & Ire: I, O, B)
Colias croceus (GEOFFROY *in* FOURCROY, 1785) Clouded Yellow (BM & Ire: I, O? B)
Gonepteryx rhamni (LINNAEUS, 1758) The Brimstone
subsp. *rhamni* (LINNAEUS, 1758)[1] (BM: R)
subsp. *gravesi* HUGGINS, 1956[2] (Ire: R)

Subfamily Pierinae DUPONCHEL, [1835]

Aporia crataegi (LINNAEUS, 1758) Black-veined White (BM: E, I or A, D)
Pieris brassicae (LINNAEUS, 1758) Large White (BM & Ire: R)
Pieris rapae (LINNAEUS, 1758) Small White (BM & Ire: R)
Pieris napi (LINNAEUS, 1758) Green-veined White
subsp. *napi* (LINNAEUS, 1758)[1] (BM: R)
subsp. *septentrionalis* VERITY, 1916[2] (BM: R)
subsp. *britannica* MÜLLER & KAUTZ, 1939[3] (Ire: R)
subsp. *thomsoni* WARREN, 1968[4] (probably = *britannica*) (BM: R)
Pontia daplidice (LINNAEUS, 1758)[+] Bath White (BM & Ire: I, O? B?)
Anthocharis cardamines (LINNAEUS, 1758) Orange-tip
subsp. *cardamines* (LINNAEUS, 1758)[1] (BM: R)
subsp. *britannica* (VERITY, 1908)[2] (BM: R)
subsp. *hibernica* (WILLIAMS, 1916)[3] (Ire: R)

Family Lycaenidae [LEACH], [1815]

Subfamily Theclinae BUTLER, 1869

Callophrys rubi (LINNAEUS, 1758) Green Hairstreak (BM & Ire: R)
Thecla betulae (LINNAEUS, 1758) Brown Hairstreak (BM & Ire: R)

Quercusia quercus (LINNAEUS, 1758) Purple Hairstreak (BM & Ire: R)
Satyrium w-album (KNOCH, 1782) White-letter Hairstreak (BM: R)
Satyrium pruni (LINNAEUS, 1758) Black Hairstreak (BM: R)

Subfamily Lycaeninae [LEACH], [1815]

Lycaena phlaeas (LINNAEUS, 1761) Small Copper
subsp. *eleus* (FABRICIUS, 1798)[1] (BM: R)
subsp. *hibernica* GOODSON, 1948[2] (Ire: R)
unnamed Scottish race (Thomson, 1980)[3] (BM: R)
Lycaena dispar (HAWORTH, 1803) Large Copper
subsp. *dispar* (HAWORTH, 1803)[1] (BM: E)
subsp. *rutilus* WERNEBERG, 1864[2] (BM & Ire: D, E)
subsp. *batavus* (OBERTHÜR, 1923)[3] (BM: D; Ire: D, E)
Lycaena virgaureae (LINNAEUS, 1758) Scarce Copper (BM: ? E)
Lycaena hippothoe (LINNAEUS, 1761) Purple-edged Copper (BM: ? E)

Subfamily Polyommatinae SWAINSON, 1827

Lampides boeticus (LINNAEUS, 1767) Long-tailed Blue (BM: I, B)
Cupido minimus (FUESSLY, 1775) Small Blue (BM & Ire: R)
Everes argiades (PALLAS, 1771) Short-tailed Blue (BM: I)
Plebejus argus (LINNAEUS, 1758) Silver-studded Blue
subsp. *argus* (LINNAEUS, 1758)[1] (BM: R)
subsp. *cretaceus* TUTT, 1909[2] (BM: R)
subsp. *masseyi* TUTT, 1909[3] (BM: E, but see Birkett, 1995)
subsp. *caernensis* THOMPSON, [1937][4] (BM: R)
Aricia agestis ([DENIS & SCHIFFERMÜLLER], 1775) Brown Argus (BM: R)
Aricia artaxerxes (FABRICIUS, 1793) Northern Brown Argus
subsp. *artaxerxes* (FABRICIUS, 1793)[1] (BM: R)
subsp. *salmacis* (STEPHENS, 1828)[2] (BM: R)
Polyommatus icarus (ROTTEMBURG, 1775) Common Blue
subsp. *icarus* (ROTTEMBURG, 1775)[1] (BM: R)
subsp. *mariscolore* (KANE, 1893)[2] (BM & Ire: R)
unique island race (Heslop Harrison, 1950b)[3] (BM: R)
Lysandra coridon (PODA, 1761) Chalk Hill Blue (BM: R)
Lysandra bellargus (ROTTEMBURG, 1775) Adonis Blue (BM: R)
Cyaniris semiargus (ROTTEMBURG, 1775) Mazarine Blue (BM: E, I)
Celastrina argiolus (LINNAEUS, 1758) Holly Blue
subsp. *argiolus*[1] (LINNAEUS, 1758) (BM: R)
subsp. *britanna*[2] (VERITY, 1919) (Ire: R)
Maculinea arion (LINNAEUS, 1758) Large Blue
subsp. *arion* (LINNAEUS, 1758)[1] (BM: D)
subsp. *eutyphron* (FRUHSTORFER, 1915)[2] (BM: E)

Subfamily Riodininae GROTE, 1895

Hamearis lucina (LINNAEUS, 1758) Duke of Burgundy Fritillary (BM: R)

Family Nymphalidae SWAINSON, 1827

Subfamily Limenitinae BEHR, 1864

Ladoga camilla (LINNAEUS, 1764) White Admiral (BM: R)

Subfamily Apaturinae BOISDUVAL, 1840

Apatura iris (LINNAEUS, 1758) Purple Emperor (BM: R)

Subfamily Nymphalinae SWAINSON, 1827

Vanessa atalanta (LINNAEUS, 1758) Red Admiral (BM & Ire: M, O, B)
Cynthia cardui (LINNAEUS, 1758) Painted Lady (BM & Ire: M, O, B)
Cynthia virginiensis (DRURY, 1773) American Painted Lady (BM & Ire: I)
Aglais urticae (LINNAEUS, 1758) Small Tortoiseshell (BM & Ire: R)
Nymphalis polychloros (LINNAEUS, 1758) Large Tortoiseshell (BM: E, I, O, B)
Nymphalis antiopa (LINNAEUS, 1758) Camberwell Beauty (BM & Ire: I, O)
Inachis io (LINNAEUS, 1758) The Peacock (BM & Ire: R)
Polygonia c-album (LINNAEUS, 1758) The Comma (BM: R)

Subfamily Argynninae DUPONCHEL, 1844

Boloria selene ([DENIS & SCHIFFERMÜLLER], 1775) Small Pearl-bordered Fritillary
subsp. *selene* ([DENIS & SCHIFFERMÜLLER], 1775)[1] (BM: R)
subsp. *insularum* (HESLOP HARRISON, 1937)[2] (BM: R)
Boloria euphrosyne (LINNAEUS, 1758) Large Pearl-bordered Fritillary (BM & Ire: R)
Argynnis lathonia (LINNAEUS, 1758) Queen of Spain Fritillary (BM & Ire: I, B?)
Argynnis adippe ([DENIS & SCHIFFERMÜLLER], 1775) High Brown Fritillary
subsp. *vulgoadippe* VERITY, 1929 (BM: R)
Argynnis aglaja (LINNAEUS, 1758) Dark Green Fritillary
subsp. *aglaja* (LINNAEUS, 1758)[1] (BM & Ire: R)
subsp. *scotica* WATKINS, 1923[2] (BM: R)
unique island race (Heslop Harrison, 1945c)[3] (BM: R)
Argynnis paphia (LINNAEUS, 1758) Silver-washed Fritillary (BM & Ire: R)

Subfamily Melitaeinae REUTER, 1896

Eurodryas aurinia (ROTTEMBURG, 1775) Marsh Fritillary
subsp. *aurinia* (ROTTEMBURG, 1775)[1] (= *anglicana* FRUHSTORFER, 1916) (BM: R)
subsp. *hibernica* (BIRCHALL, 1873)[2] (Ire: R)
subsp. *scotica* (ROBSON, 1880)[3] (BM: R)
Melitaea cinxia (LINNAEUS, 1758) Glanville Fritillary (BM: R)
Mellicta athalia (ROTTEMBURG, 1775) Heath Fritillary (BM: R)

Subfamily Satyrinae BOISDUVAL, [1833]

Pararge aegeria (LINNAEUS, 1758) Speckled Wood
subsp. *tircis* (GODART, 1821)[1] (BM & Ire: R)
subsp. *oblita* HESLOP HARRISON, 1949[2] (BM: R)
subsp. *insula* HOWARTH, 1971[3] (BM: R)
Lasiommata megera (LINNAEUS, 1767) The Wall
subsp. *megera* (LINNAEUS, 1767)[1] (BM & Ire: R)
subsp. *caledonia* VERITY, 1911[2] (BM: R)
Erebia epiphron (KNOCH, 1783) Small Mountain Ringlet
subsp. *aetherius* f. *nelamus* (Ire: ? E; Redway, 1981)
subsp. *mnemon* (HAWORTH, 1812)[1] (BM: R)
subsp. *scotica* COOKE, 1943[2] (BM: R)
Erebia aethiops (ESPER, 1777) Scotch Argus
subsp. *aethiops* (ESPER, 1777)[1] (BM: R)
subsp. *caledonia* VERITY, 1911[2] (BM: R)
Erebia ligea (LINNAEUS, 1758) Arran Brown (BM: ? E)
Melanargia galathea (LINNAEUS, 1758) Marbled White
subsp. *serena* VERITY, 1913 (BM: R)
Hipparchia semele (LINNAEUS, 1758) The Grayling
subsp. *semele* (LINNAEUS, 1758)[1] (BM: R)
subsp. *thyone* (THOMPSON, 1944)[2] (BM: R)
subsp. *scota* (VERITY, 1911)[3] (= *atlantica* (HESLOP HARRISON, 1946) (BM: R)

subsp. *clarensis* DE LATTIN, 1952[4] (Ire: R)
subsp. *hibernica* HOWARTH, 1971[5] (Ire: R)
Pyronia tithonus (LINNAEUS, 1771) The Gatekeeper
subsp. *tithonus*[1] (LINNAEUS, 1771) (BM: R)
subsp. *britanniae*[2] (VERITY, 1915) (Ire: R)
Maniola jurtina (Linnaeus, 1758) Meadow Brown
subsp. *jurtina* (LINNAEUS, 1758)[1] (BM: R)
subsp. *insularis* THOMSON, 1969[2] (BM: R)
subsp. *iernes* GRAVES, 1930[3] (Ire: R)
subsp. *cassiteridum* GRAVES, 1930[4] (BM: R)
subsp. *splendida* WHITE, 1871[5] (BM: R)
Aphantopus hyperantus (LINNAEUS, 1758) The Ringlet
subsp. *hyperantus* (LINNAEUS, 1758)[1] (BM & Ire: R)
unnamed Hebridean and Scottish race north of 56°N[2] (Dennis, 1977) (BM: R)
Coenonympha pamphilus (LINNAEUS, 1758) Small Heath
subsp. *pamphilus* (LINNAEUS, 1758)[1] (BM & Ire: R)
subsp. *rhoumensis* HESLOP HARRISON, 1948[2] (BM: R)
unique island race (Heslop Harrison, 1950b)[3] (BM: R)
Coenonympha tullia (MÜLLER, 1764) Large Heath
subsp. *scotica* STAUDINGER, 1901[1] (BM & Ire: R)
subsp. *polydama* (HAWORTH, 1803)[2] (BM: R)
subsp. *davus* (FABRICIUS, 1777)[3] (BM: R)

Subfamily Danainae BOISDUVAL, 1833

Danaus plexippus (LINNAEUS, 1758) The Monarch (BM & Ire: I)

2. LIST OF BUTTERFLIES ON BRITISH AND IRISH ISLANDS

Abbreviations and notes for the island records:
D deliberately introduced;
? identification in doubt.

No attempt has been made to distinguish breeding records from those of vagrants. The quality of data is often inadequate for this purpose and the status of insects on islands can readily change with time, especially for small islands. We feel that the onus of proof should fall on providing evidence of a breeding population (see section I.1 and Appendix 2). Superscript numbers indicate the form of so-called subspecies on islands named in the main check list; when placed in brackets the status of the form has not been formally designated. In some cases (e.g., *Lasiommata megera caledonia*) insufficient distributional details exist to designate island populations appropriately.

An asterisk indicates that the record predates 1960; some of these older records may suggest that the species no longer exists on the island or, at very least, they are in need of confirmation. The islands have been placed in regional groups and references on the butterflies for islands in each group are indicated by the number for each reference in the bibliography. Also, attention is drawn to a number of islands in each regional group which lack records and which could be usefully surveyed.

References are given to the National Grid, two letters (the first two numbers for Ireland) designating the 100 km squares, and two numbers the 10 km squares. Islands larger than one 10 km square are referenced by their most southerly points.

A. Islands off the French coast

Channel Islands

No records for Burhau, Raz and Les Casquets [WA51] or smaller islands around Alderney; Brecqhou near Sark [WV47]; Crevichon near Herm [WV38]; Lihou near Guernsey [WV27]; L'Islet, St Aubin's [WV65], La Motte [WV64], Ile de Guerdain [WV54], Les Ecréhous [WV76] and Les Minquiers [WV63] near Jersey.

ALDERNEY [WA50]: *P. machaon*[2]*; *I. podalirius**; *C. hyale*; *C. croceus*; *G. rhamni*[1]; *P. brassicae*; *P. rapae*; *P .napi*[1]; *A. cardamines*[1]; *C. rubi*; *Q. quercus**; *S. w-album*; *L. phlaeas*[1]; *P .argus*[1]; *A. agestis*; *P. icarus*[1]; *C. argiolus*[1]; *V. atalanta*; *C. cardui*; *A. urticae*; *N. polychloros*; *I. io*; *P. c-album*; *A. aglaja*[1]; *A. paphia**; *M. cinxia*; *P. aegeria*[1]; *L. megera*[1]; *H. semele*[1]; *P. tithonus*[1]; *M. jurtina*[1]; *C. pamphilus*[1].

HERM [WV38]: *C. croceus**; *G. rhamni*[1]; *P. brassicae*; *P. rapae*; *P. napi*[1]; *C. rubi*; *L. phlaeas*[1]; *P. argus*[1]; *P. icarus*[1]; *C. argiolus*[1]; *V. atalanta*; *C. cardui*; *A. urticae*; *A. aglaja*[1]; *M. cinxia*; *P. aegeria*[1]; *L. megera*[1]; *H. semele*[1]; *P. tithonus*[1]; *M. jurtina*[1]; *C. pamphilus*[1].

JETHOU [WV38]: P. brassicae*; P. rapae*; P. argus[1]; C. argiolus[1]*; A. aglaja[1]; M. jurtina[1]*; C. pamphilus[1]*.

SARK [WV47]: P. machaon[2]; I. podalirius*; C. croceus; G. rhamni[1]; P. brassicae; P. rapae; P. napi[1]; A. cardamines[1]*; C. rubi; L. phlaeas[1]; L. boeticus; P. argus[1]; A. agestis; P. icarus[1]; C. argiolus[1]; V. atalanta; C. cardui; A urticae; N. polychloros; I. io; P. c-album; B. selene[1]; A. lathonia*; A. aglaja[1]; A. paphia*; M. cinxia; P. aegeria[1]; L. megera[1]; H. semele[1]; P. tithonus[1]; M. jurtina[1]; C. pamphilus[1]; D. plexippus.

GUERNSEY [WV27]: P. machaon[2]; I. podalirius*; C. hyale; C. croceus; G. rhamni[1]; P. brassicae; P. rapae; P. napi[1]; P. daplidice*; A. cardamines[1]; C. rubi; Q. quercus; L. phlaeas[1]; H. tityrus; L. boeticus; P. argus[1]; A. agestis; P. icarus[1]; C. argiolus[1]; V. atalanta; C. cardui; A. urticae; N. polychloros; N. antiopa*; I. io; P .c-album; A. lathonia*; A. aglaja[1]; M. cinxia; P. aegeria[1]; L. megera[1]; H. semele[1]; P. tithonus[1]; M. jurtina[1]; C. pamphilus[1]*; D. plexippus*.

JERSEY [WV64]: H. morpheus; T. lineola; O. venata; P. malvae*; P. machaon[2]; L. sinapis[1]; C. hyale; C. croceus; G. rhamni[1]; G. cleopatra; A. crataegi*; P. brassicae; P. rapae; P. napi[1]; P. daplidice*; A. cardamines[1]; C. rubi; Q. quercus; S. w-album; L. phlaeas[1]; L. dispar[2]*(?); L. boeticus; E. argiades*; A. agestis; P. icarus[1]; C. semiargus*; C. argiolus[1]; L. camilla; A. iris*; V. atalanta; C. cardui; A. urticae; N. polychloros; N. antiopa; I. io; P. c-album; B. euphrosyne*; A. lathonia*; A. aglaja[1]*; A. paphia*; M. cinxia; P. aegeria[1]; L. megera[1]; H. semele[1]; P. tithonus[1]; M. jurtina[1]; A. hyperantus[1]*; C. pamphilus[1].

References
14, 39, 49, 100, 315, 345, 350, 351, 366–370, 372, 373, 378–393, 414, 432, 451, 464, 465, 528–530.

B. Islands off the British mainland

Norfolk, Suffolk, Essex and Kent
No records for Scolt Head, Norfolk [TF74], Havergate, Suffolk [TM44], Horsey [TM22], Skipper's [TM22], Mersea [TM01], Cobmarsh [TL90], Sunken [TL91], Osea [[TL90] and Northey [TL90] in Essex, nor for islands near Sheppey, such as Fowey in the Swale [TQ96].

SHEPPEY [TR06]: T. sylvestris; T. lineola; O. venata; C. croceus; P. brassicae; P. rapae; P. napi[2]; A. cardamines[2]; L. phlaeas[1]; P. icarus[1]; C. argiolus[2]; V. atalanta; C. cardui; A. urticae; N. polychloros*; I. io; P. c-album; P. aegeria[1]; L. megera[1]; P. tithonus[2]; M. jurtina[2]; C. pamphilus[1].

Reference
176.

Hampshire
HAYLING [SU79]: T. sylvestris; T. lineola; H. comma; O. venata; C. croceus; G. rhamni[1], P. brassicae; P. rapae; P. napi[2]; A. cardamines[2]; C. rubi; Q. quercus; L. phlaeas[1]; L. boeticus*; A. agestis; P. icarus[1]; C. argiolus[2]; L. camilla; V. atalanta; C. cardui; A. urticae; I. io; P. c-album; A. aglaja[1]; P. aegeria[1]; L. megera[1]; M. galathea; H. semele[1]; P. tithonus[2]; M. jurtina[2]; C. pamphilus[1].

WIGHT [SU47]: T. sylvestris; T. lineola; T. acteon*; O. venata; E. tages[1]; P. malvae; P. machaon[2]; L. sinapis[1]*; C. hyale*; C. croceus; G. rhamni[1]; A .crataegi*; P. brassicae; P. rapae; P. napi[2]; A. cardamines[2]; C. rubi; T. betulae; Q. quercus; S. w-album; L. phlaeas[1]; C. minimus; P. argus[1]; A. agestis; P. icarus[1]; L. coridon; L. bellargus; C. semiargus*; C. argiolus[2]; H. lucina; L. camilla; A. iris; V. atalanta; C. cardui; C. virginiensis*; A. urticae; N. polychloros; N. antiopa; I. io; P. c-album; B. selene[1]; B. euphrosyne; A. lathonia*; A. adippe; A. aglaja[1]; A. paphia; E. aurinia[1]*; M. cinxia; P. aegeria[1]; L. megera[1]; M. galathea; H. semele[1]; P. tithonus[2]; M. jurtina[2]; A. hyperantus[1]; C. pamphilus[1]; D. plexippus.

References
172, 336, 357, 358.

Dorset
No records for other Poole Harbour islands such as Long, Furzey, Gigger's, Green, Grove, Pergin's and Round [SU08, ST98].

BROWNSEA [SU08]: *T. sylvestris*; *O. venata*; *E. tages*[1]; *P. malvae*; *C. croceus*; *G. rhamni*[1]; *P. brassicae*; *P. rapae*; *P. napi*[2]; *A. cardamines*[2]; *C. rubi*; *Q. quercus*; *L. phlaeas*[1]; *P. argus*[1]; *P. icarus*[1]; *C. argiolus*[2]; *L. camilla*; *V. atalanta*; *C. cardui*; *A. urticae*; *I. io*; *P. c-album*; *B. selene*[1]; *A. aglaja*[1]; *A. paphia*; *P. aegeria*[1]; *L. megera*[1]; *M. galathea*; *H. semele*[1]; *P. tithonus*[2]; *M. jurtina*[2]; *A. hyperantus*[1]; *C. pamphilus*[1].

Reference
507.

South Devon and south Cornwall
No records for Burgh [SX64], Great Mewstone [SX54], Drake's [SX45], Asparagus and Gull Rock [SW61], Mullion [SW61] and Clement's [SW42].

LOOE or ST GEORGE'S [SX25]: *O. venata*; *P. brassicae*; *P. rapae*; *V. atalanta*; *A. urticae*; *P. aegeria*[1]; *M. jurtina*[2]; *A. hyperantus*[1].

ST MICHAEL'S MOUNT [SW52]: *P. brassicae*; *P. icarus*[1]; *C. argiolus*[2]; *V. atalanta*; *C. cardui*.

References
187, 491, 516.

Isles of Scilly
No records for Bishop, Gilstone, Rosevear, Annet [SV80], Mincarlo, Scilly Rock, Maiden Bower, The Minaltos, Gweal, Northwethel, Men-a-vaur [SV81], Nornour, Ragged and Toll [SV91].

ST MARTIN'S [SV91]: *C. croceus*; *G. rhamni*[1]; *P. brassicae*; *P. rapae*; *P. napi*[(2)]; *L. phlaeas*[1]; *P .icarus*[1]; *C. argiolus*[2]; *V. atalanta*; *C. cardui*; *A. urticae*; *I. io*; *P. c-album*; *P. aegeria*[3]; *L. megera*[1]; *M. jurtina*[4]; *C. pamphilus*[1].

ST MARY'S [SV81]: as for St Martins but including *D. plexippus*.

TRESCO [SV81]: *C. hyale**; *C. croceus*; *G. rhamni*[1]*; *P. brassicae*; *P. rapae*; *P. napi*[(2)]*; *L. phlaeas*[1]; *P. icarus*[1]; *C. argiolus*[2]; *V. atalanta*; *C. cardui*; *A. urticae*; *N. polychloros**; *I. io*; *P. aegeria*[3]; *M. jurtina*[4]; *D. plexippus*.

ST AGNES [SV80]: *C. croceus*; *G. rhamni*[1]; *P. brassicae*; *P. rapae*; *L. phlaeas*[1]; *P. icarus*[1]; *C. argiolus*[2]; *V. atalanta*; *C. cardui*; *A. urticae*; *I. io*; *P. c-album*; *P. aegeria*[3]; *M. jurtina*[4]; *D. plexippus*.

WHITE [SV91]: *M. jurtina*[4]*.

ST HELEN'S [SV81]: *C. croceus**; *I. io*; *M. jurtina*[4]*.

TEAN [SV91]: *P. rapae*; *P. icarus*[3]; *V. atalanta*; *C. cardui**; *P. aegeria*[3]; *M. jurtina*[4].

BRYHER [SV81]: *C. croceus*; *G. rhamni*[1]*; *P. brassicae*; *P. rapae*; *P. napi*[(2)]*; *L. phlaeas*[1]; *P. icarus*[1]; *C. argiolus*[2]; *V. atalanta*; *C. cardui*; *A. urticae*; *I. io*; *P. aegeria*[3]; *M. jurtina*[4].

SAMSON [SV81]: *P. rapae*; *P. icarus*[1]; *V. atalanta**; *C. cardui**; *M. jurtina*[4].

GREAT GANILLY [SV91]: *P. rapae*; *L. phlaeas*[1]; *P. icarus*[1]; *V. atalanta*; *C. cardui*; *I. io*; *M. jurtina*[4].

GREAT ARTHUR [SV91]: *M. jurtina*[4]*.

GUGH [SV80]: _P. aegeria_[3].

ROUND [SV91]: _C. croceus_*; _P. rapae_*; _V. atalanta_*; _C. cardui_*.

MENAWETHAN [SV91]: _P. icarus_[1]*; _M. jurtina_[4] (D).

GREAT INNISVOULS [SV91]: _L. phlaeas_[1]; _P. icarus_[1]*; _M. jurtina_[4] (D).

References
3, 4, 6–8, 26, 37, 97, 108, 109, 117, 118, 123, 130, 147, 149, 150, 170, 177, 180, 183, 186, 188, 212, 232, 329, 334, 335, 352, 405, 406, 415, 417, 449, 450, 471, 473, 490, 492, 493, 494, 568.

North Cornwall and north Devon
No records for The Brisons [SW35], Godrevy [SW54], Newland [SW98], The Mouls [SW98], Gulland Rock [SW87]; and Little Shutter Rock, Rat and Mouse around Lundy [SX14].

LUNDY [SX14]: _T. sylvestris_; _C. croceus_; _G. rhamni_[1]*; _P. brassicae_; _P. rapae_; _P. napi_[2]; _A. cardamines_[2]*; _C. rubi_; _L. phlaeas_[1]; _A. agestis_; _P. icarus_[1]; _C. argiolus_[2]; _V. atalanta_; _C. cardui_; _A. urticae_; _N. polychloros_*; _I. io_; _P. c-album_; _B. selene_[1]*; _B. euphrosyne_*; _A. aglaja_[1]; _P. aegeria_[1]; _L. megera_[1]; _H. semele_[1]; _P. tithonus_[2]; _M. jurtina_[4]; _A. hyperantus_[1]; _C. pamphilus_[1].

References
44, 47, 85, 154, 155, 342, 374, 399, 474, 521, 533, 562.

Bristol Channel
No records for Stert [SS24] or Denny [SS48].

STEEPHOLM [SS26]: _C. croceus_*; _P. brassicae_; _P. rapae_; _P. napi_[2]*; _L. phlaeas_[1]*; _P. icarus_[1]*; _C. argiolus_[2]*; _V. atalanta_*; _C. cardui_*; _A. urticae_; _I. io_*; _A. aglaja_[1]*; _P. aegeria_[1]; _L. megera_[1]; _M. jurtina_[2]*; _C. pamphilus_[1]*.

FLATHOLM [SS26]: _C. cardui_; _A. urticae_.

References
204, 401, 425.

South Wales
No records for Burry Holms [SS49], St Catherine's [SS19], St Margaret's [SN10], Thorn and Sheep [SR80], Gateholm, Mewstone and Midland [SR70], The Smalls [SR30], Ynys Bery, Yn Yscanter and Ynys Eilun [SR72], Bishops and Clerks including Carreg Rhoson [SR62], Cardigan [SN15] and Ynys Lochtyn [SN35].

PENRHYN-GWYR [SS38]: _C. croceus_*; _P. brassicae_*; _P. icarus_[1]; _V. atalanta_*; _C. cardui_*; _A. urticae_*; _I. io_*; _L. megera_[1]*; _H. semele_[1]; _M. jurtina_[2]; _C. pamphilus_[1]*.

CALDEY [SS19]: _O. venata_; _C. croceus_*; _P. brassicae_; _P. rapae_; _L. phlaeas_[1]; _C. minimus_; _P. icarus_[1]; _V. atalanta_; _A. urticae_; _I. io_; _P. aegeria_[1]; _L. megera_[1]; _H. semele_[1]*; _P. tithonus_[2]; _M. jurtina_[2]; _A. hyperantus_[1]; _C. pamphilus_[1].

SKOKHOLM [SR70]: _O. venata_; _P. malvae (?)_; _C. hyale_; _C. croceus_; _G. rhamni_[1]; _P. brassicae_; _P. rapae_; _P. napi_[2]; _A. cardamines_[2]; _L. phlaeas_[1]; _C. minimus?_; _A. agestis_; _P. icarus_[1]; _C. argiolus_[2]; _L. bellargus?_; _V. atalanta_; _C. cardui_; _A. urticae_; _I. io_; _P. c-album_; _B. selene_[1]; _A. adippe?_; _A. aglaja_[1]; _A. paphia_; _P. aegeria_[1]; _L. megera_[1]; _M. galathea_; _H. semele_[1]; _P. tithonus_[2]; _M. jurtina_[2]; _A. hyperantus_[1]; _C. pamphilus_[1].

SKOMER [SR70]: _T. sylvestris_; _O. venata_; _C. croceus_; _P. brassicae_; _P. rapae_; _P. napi_[2]; _A. cardamines_[2]*;

L. phlaeas[1]; *P. icarus*[1]; *V. atalanta*; *C. cardui*; *A. urticae*; *I. io*; *P. c-album**; *B. selene*[1]*; *A. aglaja*[1]; *L. megera*[1]; *H. semele*[1]; *P. tithonus*[2]; *M. jurtina*[2]; *A. hyperantus*[1]; *C. pamphilus*[1]; *D. plexippus*.

RAMSEY [SR72]: *O. venata*; *P. brassicae*; *P. rapae*; *P. napi*[2]; *L. phlaeas*[1]; *P. icarus*[1]; *V. atalanta*; *C. cardui*; *A. urticae*; *I. io*; *A. aglaja*[1]; *L. megera*[1]; *H. semele*[1]; *P. tithonus*[2]; *M. jurtina*[2].

GRASSHOLM [SR51]: *P. brassicae*; *P. rapae**; *V. atalanta**; *C. cardui*; *A. urticae**.

References
40, 99, 214, 230, 231, 237, 475.

North Wales
No records from Shell [SH52], St Tudwall's Islands [SH32], Ynys Gwylan [Sh12], Llanddwyn [SH36], North Stack and South Stack [SH28], The Skerries [SH29], Mouse Islands [SH49], Ynys Dulas [SH59] and Ynys Moelfre [SH58].

BARDSEY [SH12]: *C. croceus*; *P. brassicae*; *P. rapae*; *P. napi*[(2)]; *A. cardamines*[2]; *L. phlaeas*[1]; *P. icarus*[1]; *V. atalanta*; *C. cardui*; *A. urticae*; *I. io*; *A. paphia**; *L. megera*[1]; *H. semele*[1]; *P. tithonus*[2]; *M. jurtina*[2]; *C. pamphilus*[1].

ANGLESEY [SH46]: *T. sylvestris*; *O. venata*; *E. tages*[1]; *P. malvae*; *C. croceus*; *G. rhamni*[1]; *P. brassicae*; *P. rapae*; *P. napi*[(2)]; *A. cardamines*[2]; *C. rubi*; *Q. quercus*; *S. w-album*; *L. phlaeas*[1]; *P. argus*[1]; *A. agestis*; *P. icarus*[1]; *C. argiolus*[2]; *V. atalanta*; *C. cardui*; *A. urticae*; *N. antiopa*; *I. io*; *P. c-album*; *B. selene*[1]; *B. euphrosyne*; *A. adippe*; *A. aglaja*[1]; *A. paphia**; *E. aurinia*[1]; *P. aegeria*[1]; *L. megera*[1]; *H. semele*[1]; *P. tithonus*[2]; *M. jurtina*[2]; *A. hyperantus*[1]; *C. pamphilus*[1].

HOLY [SH27]: *O. venata*; *C. croceus*; *P. brassicae*; *P. rapae*; *P. napi*[(2)]; *A. cardamines*[2]; *L. phlaeas*[1]; *P. argus*[1]; *P. icarus*[1]; *V. atalanta*; *C. cardui*; *A. urticae*; *I. io*; *B. selene*[1]; *A. aglaja*[1]; *E. aurinia*[1]; *L. megera*[1]; *H. semele*[1]; *P. tithonus*[2]; *M. jurtina*[2]; *A. hyperantus*[1]; *C. pamphilus*[1].

PUFFIN [SH68]: *C. croceus**; *P. brassicae**; *P. rapae**; *P. napi*[(2)]*; *V. atalanta**; *C. cardui*; *A. urticae**; *I. io**; *L. megera*[1]*; *H. semele*[1]*; *M .jurtina*[2]*.

References
12, 90, 128, 132, 216, 408, 418, 554.

Irish Sea and north-west England
No records for Calf Islands, Kitterland, The Burroo, The Stack, St Patrick's [NX28] and St Michael's [NX26] in the vicinity of Man; nor from Little Hilbre and Little Eye [SJ18], Foulney, Roa, Piel and Whalney [SD26].

HILBRE [SJ18]: *T. sylvestris*; *O. venata*; *C. croceus*; *P. brassicae*; *P. rapae*; *P. napi*[2]; *L. phlaeas*[1]; *P. icarus*[1]; *C. argiolus*[2]; *V. atalanta*; *C. cardui*; *A. urticae*; *I. io*; *P. c-album*; *P. aegeria*[1]; *L. megera*[1]; *H. semele*[1]; *P. tithonus*[2]; *M. jurtina*[2]; *C. pamphilus*[1].

MAN [NX16]: *O. venata**; *C. croceus*; *G. rhamni*[1](+D); *P. brassicae*; *P. rapae*; *P. napi*[(2)]; *A. cardamines*[3]; *L. phlaeas*[1]; *C. minimus*(?); *P. icarus*[1]; *C. argiolus*[2]; *V. atalanta*; *C. cardui*; *A. urticae*; *N. antiopa*; *I. io*; *P. c-album* (D); *B. selene*[1]*; *B. euphrosyne**; *A. aglaja*[1]; *E. aurinia*[1]*(?); *P. aegeria*[1](+D); *L. megera*[1]; *E. aethiops*[1]*(?); *H. semele*[1]; *P. tithonus*[2]*; *M. jurtina*[3]; *A. hyperantus*[1]*; *C. pamphilus*[1]; *C. tullia*[(2)]*; *D. plexippus*.

CALF OF MAN [NX16]: *C. croceus*; *P. brassicae*; *P. rapae*; *P. napi*[(2)]; *A. cardamines*[3]; *L. phlaeas*[1]; *P. icarus*[1]; *V. atalanta*; *C. cardui*; *A. urticae*; *I. io*; *L. megera*[1]; *H. semele*[1]; *M. jurtina*[3]; *C. pamphilus*[1].

CHICKEN ROCK [NX16]: *C. cardui*.

References
5, 36, 93, 106, 107, 124, 233–235, 397, 398, 457, 509, 518.

South-west Scotland and Firth of Clyde
No records for Rough and Hestan [NX85], Little Ross [NX64], islands of Fleet [NX54], Inner and Outer Eileans [NS15].

AILSA CRAIG [NX09]: *E. tages*(?); *P. brassicae*; *P. rapae*; *P. napi*[4]; *C. rubi*; *L. phlaeas*[(3)]; *A. artaxerxes*[1]* (recorded as *A. agestis*); *P. icarus*[1]; *V. atalanta*; *C. cardui*; *A. urticae*; *I. io**; *B. selene*[1](?); *A. aglaja*[1]*; *E. aethiops*[2]; *H. semele*[1]; *M. jurtina*[(5)]*; *C. pamphilus*[1]*; *C. tullia*[(2)]*.

LADY ISLE [NS22]: *P. brassicae*; *P. rapae*; *P. napi*[4]; *L. phlaeas*[(3)]; *P. icarus*[1]; *V. atalanta*; *C. cardui*; *A. urticae*; *I. io*; *H. semele*[1]; *M. jurtina*[(5)]; *C. pamphilus*[1].

HORSE [NS24]: *P. brassicae*; *P. rapae*; *P. napi*[4]; *P. icarus*[1]; *V. atalanta*; *A. urticae*; *I. io*; *H. semele*[1]; *M. jurtina*[(5)].

LITTLE CUMBRAE [NS15]: *P. brassicae*; *P. rapae*; *P. napi*[4]; *L. phlaeas*[1]; *P. icarus*[1]; *V. atalanta*; *C. cardui*; *A. urticae*; *I. io*; *B. selene*[1]; *A. aglaja*[1]; *E. aethiops*[2]; *H. semele*[1]; *M. jurtina*[(5)]; *C. pamphilus*[1]; *C. tullia*[1].

GREAT CUMBRAE [NS15]: *P. brassicae*; *P. rapae*; *P. napi*[4]; *C. rubi*; *L. phlaeas*[1]; *P. icarus*[1]; *V. atalanta*; *C. cardui*; *A. urticae*; *I. io*; *B. selene*[1]; *A. aglaja*[1]; *E. aethiops*[2]; *H. semele*[1]; *M. jurtina*[(5)]; *C. pamphilus*[1].

ARRAN [NR92]: *C. croceus*; *P. brassicae*; *P. rapae*; *P. napi*[4]; *C. rubi*; *Q. quercus*; *L. phlaeas*[3]; *C. minimus**; *P. icarus*[1]; *V. atalanta*; *C. cardui*; *A. urticae*; *I. io*; *B. selene*[1]; *B. euphrosyne**; *A. aglaja*[1]; *P. aegeria*[2]; *L. megera*; *E. aethiops*[2]; *H. semele*[1]; *M. jurtina*[5]; *A. hyperantus*[(2)]*; *C. pamphilus*[1]; *C. tullia*[1]*.

HOLY [NS02]: *P. brassicae*; *P. rapae*; *P. napi*[4]; *L. phlaeas*[3]; *P. icarus*[1]; *V. atalanta*; *A. urticae*; *I. io*; *B. selene*[1]; *A. aglaja*[1]; *E. aethiops*[2]; *H. semele*[1]; *M. jurtina*[5]; *C. pamphilus*[1]; *C. tullia*[1]*.

PLADDA [NS01]: *P. brassicae*; *P. rapae*; *P. napi*[4]; *V. atalanta*; *A. urticae*; *H. semele*[1]; *M. jurtina*[5]; *C. pamphilus*[1].

INCHMARNOCK [NS05]: *P. brassicae*; *P. rapae*; *P. napi*[4]; *L. phlaeas*[3]; *P. icarus*[1]; *V. atalanta*; *A. urticae*; *I. io*; *B. selene*[1]; *A. aglaja*[1]; *H. semele*[1]; *M. jurtina*[5]; *C. pamphilus*[1].

BUTE [NS05]: *O. venata**(?); *C. croceus*; *P. brassicae*; *P. rapae*; *P. napi*[4]; *C. rubi*; *L. phlaeas*[3]; *P. icarus*[1]; *V. atalanta*; *C. cardui*; *A. urticae*; *I. io*; *B. selene*[1]; *A. aglaja*[1]; *L. megera*[(2)]; *E. aethiops*[2]; *E. ligea**(?); *H. semele*[1]; *M. jurtina*[5]; *C. pamphilus*[1].

SANDA [NR70]: *P. brassicae*; *P. rapae*; *P. napi*[4]; *L. phlaeas*[3]; *P. icarus*[1]; *V. atalanta*; *A. urticae*; *I. io*; *E. aethiops*[2]; *H. semele*[1]; *M. jurtina*[5]; *C. pamphilus*[1].

GLUNIMORE [NR70]: *P. rapae*; *A. urticae*.

SHEEP [NR70]: *P. rapae*; *P. napi*[4]; *A. urticae*; *H. semele*[1]; *M. jurtina*[5].

DAVAAR [NR71]: *P. brassicae*; *P. rapae*; *P. napi*[4]; *L. phlaeas*[3]; *P. icarus*[1]; *V. atalanta*; *A. urticae*; *I. io*; *H. semele*[1]; *M. jurtina*[5]; *C. pamphilus*[1]; *C. tullia*[1].

SGAT MOR [NR96]: *P. rapae*; *A. urticae*.

GLAS EILEAN [NR98]: *P. brassicae*; *P. rapae*; *P. napi*[4]; *A. urticae*; *M. jurtina*[(5)].

MINARD [NR99]: *P. rapae*; *V. atalanta*; *A. urticae*; *M. jurtina*[5].

BURNT ISLANDS [NS07]: *P. brassicae*; *P. rapae*; *P. napi*[4]; *A. urticae*; *H. semele*[1]; *M. jurtina*[5]; *C. pamphilus*[1].

References
102, 163, 168, 191–198, 211, 217, 394, 413, 448, 485, 486, 487, 511, 543, 548, 553, 555.

Inner Hebrides: Strathclyde
No records for Texa [NR34], Am Fraoch Eilean and Brosdale [NR46], Small Isles [NR56], Eilean Mór [NR67], Shuna [NR70], Dubh Artach [NR10], Torran Rocks [NR21], Erraid [NR21], Erisgeir [NR33], Inch Kenneth and Eorsa [NR43], Little Colonsay [NR33], Calve, Oronsay [NM55] and Carna [NM65].

GIGHA [NR64]: *P. brassicae*; *P. rapae*; *P. napi*[4]; *A. cardamines*[2]; *C. rubi*; *L. phlaeas*[3]; *P. icarus*[1]; *V. atalanta**; *C. cardui**; *A. urticae*; *I. io*; *M. jurtina*[5]; *C. pamphilus*[1]*.

CARA [NR64]: *P. rapae*; *P. napi*[4]; *L. phlaeas*[3]; *P. icarus*[1]; *V. atalanta**; *C. cardui*; *A. urticae*; *I. io**; *H. semele*[1]; *M. jurtina*[5]; *C. pamphilus*[1].

ISLAY [NR34]: *C. croceus**; *P. brassicae*; *P. rapae*; *P. napi*[4]; *A. cardamines*[2]; *C. rubi*; *L. phlaeas*[3]; *P. argus*[1](?); *A. artaxerxes*[1](?); *P. icarus*[1]; *V. atalanta*; *C. cardui*; *A. urticae*; *I. io*; *A. adippe** (? mistaken for *A. aglaja*); *A. aglaja*[1]; *E. aurinia*[3]; *P. aegeria*[2]; *L. megera*[(2)]; *H. semele*[3]; *M. jurtina*[5]; *A. hyperantus*[2]; *C. pamphilus*[1]; *C. tullia*[1].

JURA [NR56]: *P. brassicae*; *P. rapae*; *P. napi*[4]; *C. rubi*; *L. phlaeas*[3]; *P. icarus*[1]; *V. atalanta*; *C. cardui*; *A. urticae*; *I. io*; *B. selene*[(2)]; *A. aglaja*[1]; *E. aurinia*[3]; *P. aegeria*[2]; *L. megera*; *H. semele*[(3)]; *M. jurtina*[5]; *A. hyperantus*[(2)]; *C. pamphilus*[1]; *C. tullia*[1].

COLONSAY [NR38]: *P. brassicae*; *P. rapae*; *P. napi*[4]; *C. rubi*; *Q. quercus*; *L. phlaeas*[3]; *P. icarus*[1]; *V. atalanta*; *C. cardui*; *A. urticae*; *I. io*; *A. aglaja*[1]; *E. aurinia*[3]; *H. semele*[(3)]; *M. jurtina*[5]; *A. hyperantus*[(2)]; *C. pamphilus*[1]; *C. tullia*[1]*.

ORONSAY [NR38]: *P. napi*[4]; *L. phlaeas*[(3)]; *V. atalanta*.

SCARBA [NM36]: *P. icarus*[1].

GARVELLACHS [NM60 & 61]: *P. napi*[4]; *L. phlaeas*[3]; *P. icarus*[1]; *A. aglaja*[1]; *M. jurtina*[5]; *A. hyperantus*[(2)].

LUING [NM70]: *P. napi*[4]; *P. icarus*[1].

LUNGA [NM70]: *P. brassicae*; *P. napi*[4]; *C. rubi*; *L. phlaeas*[3]; *P. icarus*[1]; *B. selene*[(2)]; *A. aglaja*[1]; *E. aethiops*[2]; *H. semele*[(3)]; *M. jurtina*[5]; *C. pamphilus*[1].

SEIL [NM71]: *P. brassicae*; *P. rapae*; *P. napi*[4]; *L. phlaeas*[3]; *P. icarus*[1]; *V. atalanta*; *C. cardui*; *A. urticae*; *I. io*; *P. aegeria*[2]; *E. aethiops*[2]; *H. semele*[3]; *M. jurtina*[5]; *A. hyperantus*[2]; *C. pamphilus*[1].

EASDALE [NM71]: *P. rapae*; *P. napi*[4]; *L. phlaeas*[3]; *P. icarus*[1]; *V. atalanta*; *A. urticae*; *E. aethiops*[2]; *H. semele*[(3)]; *M. jurtina*[5]; *C. pamphilus*[1].

KERRERA [NM82]: *P. brassicae*; *P. rapae*; *P. napi*[4]; *C. rubi*; *C. minimus**; *P. icarus*[1]; *V. atalanta*; *C. cardui*; *A. urticae*; *I. io*; *B. selene*[(2)]; *A. aglaja*[1]; *E. aethiops*[2]; *H. semele*[(3)]; *M. jurtina*[5]; *C. pamphilus*[1]; *C. tullia*[1].

LISMORE [NM73]: *P. napi*[4]; *L. phlaeas*[3]; *P. icarus*[1]; *V. atalanta*; *A. urticae*; *I. io*; *B. selene*[(2)]; *E. aurinia*[3]; *H. semele*[(3)]; *M. jurtina*[5]; *C. pamphilus*[1]; *C. tullia*[1].

MULL [NM31]: *O. venata*; *P. brassicae*; *P. rapae*; *P. napi*[4]; *C. rubi*; *L. phlaeas*[3]; *P. icarus*[(2)]; *V. atalanta*; *C. cardui*; *A. urticae*; *I. io*; *B. selene*[(2)]; *B. euphrosyne*; *A. aglaja*[2]; *E. aurinia*[3]; *P. aegeria*[2]; *E. aethiops*[2]; *H. semele*[3]; *M. jurtina*[5]; *C. pamphilus*[1]; *C. tullia*[1].

TRESHNISH ISLES (largest LUNGA) [NM23]: *P. icarus*; *C. cardui*.

IONA [NM22]: *P. brassicae*; *P. rapae*; *P. napi*[4]; *C. rubi*; *P. icarus*[2]; *V. atalanta*; *C. cardui*; *A. urticae*; *I. io*; *H. semele*[3]; *M. jurtina*[5]; *C. pamphilus*[1].

SOA [NM21]: *P. napi*[4]; *H. semele*[(3)]*.

ULVA [NM43]: *P. brassicae*; *P. napi*[4]; *C. rubi*; *L. phlaeas*[3]; *P. icarus*[(2)]; *V. atalanta**; *A. urticae*; *I. io**; *B. selene*[(2)]; *A. aglaja*[(2)]*; *E. aurinia*[3]*; *P. aegeria*[2]; *E. aethiops*[2]; *H. semele*[(3)]*; *M. jurtina*[5]; *C. pamphilus*[1]*.

STAFFA [NM33]: *P. icarus*[(2)]; *A. urticae*; *H. semele*[(3)]; *M. jurtina*[5]; *C. pamphilus*[1]; *C. tullia*[1].

TIREE [NL93]: *P. brassicae*; *P. rapae*; *P. napi*[4]; *P. icarus*[2]; *V. atalanta*; *C. cardui*; *A. urticae*; *I. io*; *E. aurinia*[3]*; *H. semele*[(3)]; *M. jurtina*[5]; *C. pamphilus*[1]*; *C. tullia*[1]*.

GUNNA [NM05]: *P. brassicae**; *P. rapae**; *P. napi*[4]*; *P. icarus*[2]*; *V. atalanta**; *C. cardui**; *A. urticae**; *E. aurinia*[3]*; *H. semele*[3]; *M. jurtina*[5]*; *C. pamphilus*[1]*.

COLL [NM15]: *C. hyale**; *C. croceus*; *P. brassicae*; *P. rapae*; *P. napi*[4]; *C. rubi*; *P. icarus*[2]; *V. atalanta*; *C. cardui*; *A. urticae*; *I. io*; *B. selene*[2]; *A. aglaja*[(2)]; *H. semele*[3]; *M. jurtina*[5]; *C. pamphilus*[1]*; *C. tullia*[1].

References
1, 21, 49, 148, 151, 158, 218, 238, 242, 248, 250–252, 255-257, 264, 270, 280, 281, 286, 287, 289–291, 294–296, 300, 309, 361, 407, 410, 453, 454, 495, 508, 511, 524, 542, 553, 556–558.

Inner Hebrides: Highland
No records for Ascrib Islands [NG36], Eilean Trodday [NG47], Ornsay [NG71], Isay [NG25] and Harlosh [NG23].

MUCK [NM47]: *P. brassicae*; *P. napi*[4]; *P. icarus*[2]; *V. atalanta**; *A. urticae*; *I. io**; *A. aglaja*[(2)]; *P. aegeria*[2]; *H. semele*[(3)]; *M. jurtina*[5]; *C. pamphilus*[2].

EILEAN NAN EACH [NM38]: *P. napi*[4]*; *A. urticae**.

EIGG [NM48]: *C. croceus**; *P. brassicae**; *P. rapae**; *P. napi*[4]*; *C. rubi*; *P. icarus*[2]*; *V. atalanta*; *A. urticae**; *I. io**; *A. aglaja*[2]; *P. aegeria*[2]; *H. semele*[3]; *M. jurtina*[5]; *A. hyperantus*[(2)]*; *C. pamphilus*[2]*; *C. tullia*[1].

RHUM [NM39]: *P. brassicae*; *P. rapae*; *P. napi*[4]; *C. rubi*; *P. icarus*[2]; *V. atalanta*; *C. cardui*; *A. urticae*; *I. io*; *B. selene*[2]; *B. euphrosyne**; *A. aglaja*[2]; *E. aurinia*[3]*; *P. aegeria*[2]; *E. aethiops*[2]*; *H. semele*[3]; *M. jurtina*[5]; *C. pamphilus*[2]; *C. tullia*[1].

SANDAY [NG20]: *P. brassicae*; *P. napi*[4]; *C. rubi*; *P. icarus*[2]; *V. atalanta*; *C. cardui*; *A. urticae**; *B. selene*[(2)]; *A. aglaja*[(2)]; *P. aegeria*[2]; *H. semele*[3]; *M. jurtina*[5]; *C. pamphilus*[(2)].

CANNA [NG20]: *C. croceus*; *P. brassicae*; *P. rapae*; *P. napi*[4]; *C. rubi*; *P. icarus*[2]; *V. atalanta*; *C. cardui*; *A. urticae*; *I. io*; *B. selene*[(2)]; *A. aglaja*[2]; *P. aegeria*[2]; *H. semele*[3]; *M. jurtina*[5]; *C. pamphilus*[2]; *C. tullia*[1]*.

HEISKER [NM19]: *P. napi*[4]*; *P. icarus*[2]*; *V. atalanta**; *A. urticae**; *I. io**; *M. jurtina*[5]*.

SOAY [NG41]: *P. brassicae**; *P. napi*[4]; *P. icarus*[2]*; *A. urticae**; *I. io**; *B. selene*[2]*; *A. aglaja*[2]*; *H. semele*[3]*; *M. jurtina*[5]*; *C. pamphilus*[2]*; *C. tullia*[1]*.

WIAY [NG23]: *E. aethiops*[2].

SKYE [NM59]: *P. brassicae*; *P. rapae*; *P. napi*[4]; *A. cardamines*[2]; *C. rubi*; *P. icarus*[2]; *V. atalanta*; *C. cardui*; *A. urticae*; *N. antiopa**; *I. io*; *B. selene*[2]; *B. euphrosyne**; *A. aglaja*[2]; *P. aegeria*[2]; *E. aethiops*[2]; *H. semele*[(3)]; *M. jurtina*[5]; *A. hyperantus*[(2)]*; *C. pamphilus*[(2)]; *C. tullia*[1]; *D. plexippus**.

SCALPAY [NG62]: *P. brassicae**; *P. rapae**; *P. napi*[4]; *C. rubi**; *P. icarus*[2]; *V. atalanta**; *C. cardui**; *A. urticae**; *B. selene*[2]; *B. euphrosyne**; *A. aglaja*[2]*; *E. aethiops*[2]; *H. semele*[3]*; *M. jurtina*[5]; *C. pamphilus*[2]; *C. tullia*[1].

LONGAY [NG63]: *P. brassicae**; *P. napi*[4]*; *C. rubi**; *P. icarus*[2]*; *V. atalanta**; *A. urticae*; *B. selene*[(2)]*; *A. aglaja*[(2)]*; *E. aethiops*[2]*; *H. semele*[(3)]*; *C. pamphilus*[2]*; *C. tullia*[1]*.

PABAY [NG62]: *P. brassicae**; *P. napi*[4]*; *P. icarus*[2]*; *A. urticae**; *M. jurtina*[5]*; *C. pamphilus*[2]*; *C. tullia*[1]*.

RAASAY [NG53]: *P. brassicae*; *P. rapae*; *P. napi*[4]; *C. rubi**; *P. icarus*[2]; *V. atalanta**; *C. cardui**; *A. urticae*; *I. io**; *B. selene*[2]*; *B. euphrosyne**; *A. aglaja*[2]*; *P. aegeria*[2]; *E. aethiops*[2]*; *H. semele*[3]; *M. jurtina*[5]; *C. pamphilus*[2]; *C. tullia*[1].

FLADDAY [NG55]: *P. brassicae**; *P. napi*[4]*; *P. icarus*[(2)]*; *A. urticae**; *M. jurtina*[5]*; *C. pamphilus*[2]*; *C. tullia*[1]*.

SOUTH RONA [NG65]: *P. brassicae*; *P. napi*[4]; *P. icarus*[2]; *V. atalanta*; *C. cardui*; *A. urticae*; *B. selene*[(2)]*; *A. aglaja*[2]; *E. aethiops*[2]*; *H. semele*[3]*; *M. jurtina*[5]; *C. pamphilus*[2]; *C. tullia*[1]*.

CROWLIN ISLES [NG63] (EILEAN MÓR and EILEAN MEADLIONACH): *A. urticae*.

References
33, 41, 53–81, 115, 162, 169, 178, 190, 209, 239–245, 259, 263–266, 270–274, 276, 278, 280, 282, 284, 285, 289–292, 294–297, 300, 305, 309, 310, 312, 313, 338, 349, 355, 416, 426, 508, 511, 553, 564, 565.

North-west and north coast of Scotland
No records for Ewe [NG88] and Gruinard [NG99], nor for several islands in the Summer Islands group including Isle Martin [NH09], Horse and most Carn Islands [NC00], and Isle Ristol [NB91]; no records for Soyea [NC02], Oldany [NC03], Eilean Hoan [NC46], Eilean Choraidh [NC45], Rabbit Islands, Eilean Ròn and Neave [NC66].

LONGA [NG77]: *M. jurtina*[5].

EILEAN FURADH MÓR [NG79]: *P. icarus*[(2)]; *H. semele*[(3)].

SUMMER ISLANDS (not distinguished) [NB90]: *E. aethiops*[2]; *H. semele*[(3)]; *M. jurtina*[5].

TANERA BEG [NB90]: *P. brassicae*; *P. napi*[4]; *P. icarus*[(2)]; *E. aethiops*[2]; *M. jurtina*[5].

TANERA MÓR [NB90]: *P. brassicae*; *P. napi*[4]; *P. icarus*[(2)]; *E. aethiops*[2]; *H. semele*[(3)]; *M. jurtina*[5].

PRIEST [NB90]: *A. urticae*.

CARN NAN SGEIR [NC00]: _C. cardui._

HANDA [NC14]: _P. brassicae_; _P. rapae_; _P. napi_[4]; _P. icarus_[(2)]; _V. atalanta_; _C. cardui_; _A. urticae_; _I. io_; _E. aethiops_[2]; _M. jurtina_[5]; _C. pamphilus_[(2)]; _C. tullia_[1].

References
246, 511, 553.

Outer Hebrides
No records for Stack [NF70], Calvay [NF81], Oronsay [NF87], Boreray [NF88], Hermetray [NF97], Mealastra [NF92] and Eilean Chalium Chille [NB32].

BERNERAY [NL57]: _P. brassicae_*; _P. icarus_[2]*; _A. aglaja_[(2)]*; _H. semele_[3]*; _M. jurtina_[5]*; _C. tullia_[1].

MINGULAY [NL58]: _P. icarus_[2]*; _A. aglaja_[(2)]*; _H. semele_[3]*; _M. jurtina_[5]*; _C. pamphilus_[(2)]*; _C. tullia_[1]*.

PABBAY [NL68]: _P. icarus_[3]*; _A. aglaja_[2]*; _H. semele_[3]*.

SANDRAY [NL69]: _P. icarus_[2]*; _C. cardui_*; _A. aglaja_[2]*; _H. semele_[3]*.

VATERSAY [NL69]: _P. brassicae_*; _P. icarus_[2]; _A. aglaja_[2]*; _H. semele_[3]*; _M. jurtina_[5].

MULDOANICH [NL69]: _P. icarus_[3]*; _A. aglaja_[1]*; _M. jurtina_[5]*.

BARRA [NL69]: _C. croceus_*; _P. brassicae_; _P. rapae_; _P. napi_[4]; _A. cardamines_[2](?); _P. icarus_[2]; _V. atalanta_; _C. cardui_; _A. urticae_; _I. io_*; _A. aglaja_[2]; _H. semele_[3]; _M. jurtina_[5]; _C. tullia_[1]*.

FLODDAY [NL69]: _P. icarus_[3]*; _A. aglaja_[3]*; _H. semele_[3]*.

UINESSAN [NL69]: _A. urticae_*; _A. aglaja_[2]*; _H. semele_[3]*.

GIGHAY [NF70]: _P. napi_[4]; _P. icarus_[2]; _A. urticae_; _A. aglaja_[2]; _M. jurtina_[5].

HELLISAY [NF70]: _A. aglaja_[2]*.

FUDAY [NF70]: _P. icarus_[2]*; _A. aglaja_[2]*; _H. semele_[3]*.

FIARAY [NF71]: _P. icarus_[3]*.

ERISKAY [NF70]: _C. croceus_*; _P. brassicae_; _P. napi_[4]*; _P. icarus_[2]; _V. atalanta_*; _C. cardui_; _A. urticae_; _A. aglaja_[2]*; _H. semele_[3]*; _M. jurtina_[5]*; _C. pamphilus_[2]*.

SOUTH UIST [NF81]: _P. brassicae_; _P. rapae_; _P. napi_[4]; _A. cardamines_[2]; _P. icarus_[2,3]; _V. atalanta_; _C. cardui_; _A. urticae_; _I. io_; _A. aglaja_[2]; _H. semele_[3]; _M. jurtina_[5]; _C. pamphilus_[2,3]; _C. tullia_[1].

BENBECULA [NF84]: _P. brassicae_; _P. rapae_; _P. napi_[4]; _P. icarus_[2]; _V. atalanta_; _C. cardui_; _A. urticae_; _A. aglaja_[2]; _M. jurtina_[5]; _C. pamphilus_[(2)]; _C. tullia_[1]; _D. plexippus_*.

CALAVAY [NF85]: _P. icarus_[2].

WIAY [NF84]: _C. pamphilus_[(2)].

MONACH ISLANDS [NF66]: _P. napi_[4]; _P. icarus_[2]; _A. urticae_; _M. jurtina_[5]*.

RONAY [NF85]: _P. icarus_[2]*; _M. jurtina_[5]*.

GRIMSAY [NF85]: M. *jurtina*[5]*.

NORTH UIST [NF85]: P. *brassicae*; P. *rapae*; P. *napi*[4]; P. *icarus*[2]; V. *atalanta*; C. *cardui*; A. *urticae*; I. *io*; M. *jurtina*[5]; C. *pamphilus*[2]; C. *tullia*[1].

BALESHARE [NF75]: P. *icarus*[2]*; A. *urticae**; M. *jurtina*[5]*.

BERNERAY [NF98]: P. *brassicae**; P. *icarus*[2]*; M. *jurtina*[5]; C. *tullia*[1]*.

PABBAY [NF88]: P. *icarus*[2]*; H. *semele*[3]*.

SHILLAY [NF89]: P. *icarus*[2]*; M. *jurtina*[5]*.

ENSAY [NF98]: P. *napi*[4]*; P. *icarus*[2]*; V. *atalanta*; C. *cardui*; A. *urticae**; M. *jurtina*[5]*.

KILLEGRAY [NF98]: P. *icarus*[2,3]*; C. *cardui**; M. *jurtina*[5]*.

TARANSAY [NF99]: P. *icarus*[2]*; A. *urticae**; M. *jurtina*[5]; C. *tullia*[1]*.

HARRIS-LEWIS [NG08]: C. *croceus**; P. *brassicae*; P. *rapae*; P. *napi*[4]; P. *icarus*[2]; V. *atalanta*; C. *cardui*; A. *urticae*; I. *io*; P. *c-album*; H. *semele*[3]; M. *jurtina*[5]; C. *pamphilus*[(2)]; C. *tullia*[1].

SCARP [NA91]: P. *brassicae**; P. *icarus*[2]*; M. *jurtina*[5]*.

SCOTASAY [NB19]: P. *icarus*[2]*; M. *jurtina*[5]*.

GREAT BERNERA [NB13]: P. *icarus**[2]; V. *atalanta**.

LITTLE BERNERA [NB14]: P. *icarus*[2]*.

SHIANT ISLANDS [NG49]; P. *napi*[4]*; P. *icarus*[2,3]*; M. *jurtina*[5]*.

ST KILDA (HIRTA, SOAY & BORERAY) [NF09]: C. *croceus*; V. *atalanta**; C. *cardui*; A. *urticae*; C. *pamphilus*[(2)]*.

NORTH RONA [HW83]: A. *urticae**.

References
28, 35, 51–53, 56–58, 87, 103, 112–114, 116, 119, 121, 157, 166, 181, 184, 207, 242, 244, 247, 249, 253–255, 258, 261–265, 267–269, 271–273, 275, 277, 280, 282, 283, 287–291, 293, 294, 297–302, 304, 306–309, 312, 313, 360, 365, 396, 403, 463, 481, 482, 508, 511, 514, 517, 532, 539, 545, 552, 553.

Orkney Islands
No records for Stroma [ND37], Swona [ND38], Pentland Skerries [ND47], Cava, Rysa Little, Fara and Flotta [ND39], Brough of Birsay [HY22], Eynhallow [HY32], Wyre [HY42], Copinsay [HY60], Egilsay and Gairsay [HY42], Eday [HY52], Calf of Eday [HY53], Faray [HY53], Stronsay and Papa Stronsay [HY62] and Auskerry [HY61].

HOY [HW83]: P. *brassicae*; P. *rapae*; P. *napi*[4]; L. *phlaeas*[3]*; P. *icarus*[2]; V. *atalanta*; C. *cardui*; A. *urticae*; N. *antiopa**; I. *io*; A. *aglaja*[2]; M. *jurtina*[5]; C. *tullia*[1].

SOUTH RONALDSAY [ND48]: P. *brassicae*; P. *rapae*; P. *napi*[4]; P. *icarus*[2]; V. *atalanta*; C. *cardui*; A. *urticae*; A. *aglaja*[2]; M. *jurtina*[5]; C. *tullia*[1].

BURRAY [ND49]: *P. brassicae*; *P. napi*[4]; *P. icarus*[2]; *C. cardui*; *A. urticae*; *A. aglaja*[2]; *M. jurtina*[5].

GRAEMSAY [HY20]: *A. urticae*.

MAINLAND ORKNEY [ND50]: *C. croceus**; *P. brassicae*; *P. rapae*; *P. napi*[4]; *P. icarus*[2]; *V. atalanta*; *C. cardui*; *A. urticae*; *I. io*; *A. aglaja*[2]; *M. jurtina*[5]; *C. tullia*[1].

SHAPINSAY [HY51]: *P. brassicae*; *P. icarus*[2]; *V. atalanta*.

ROUSAY [HY42]: *P. brassicae*; *P. icarus*[2]; *C. cardui*; *A. urticae*.

WESTRAY [HY43]: *P. brassicae*; *C. cardui*; *A. urticae*.

SANDAY [HY63]: *P. brassicae*; *P. rapae*; *P. napi*[4]; *P. icarus*[2]; *V. atalanta*; *A. urticae**.

NORTH RONALDSAY [HY75]: *P. brassicae*; *P. rapae*; *P. napi*[4]; *V. atalanta*; *C. cardui*; *A. urticae*; *M. jurtina*[5].

References
38, 46, 94, 157, 201, 203, 215, 260, 319–324, 328, 375–377, 400, 455, 462, 482, 488, 496, 508, 511, 513, 514, 547, 552, 553, 569, 570, 571.

Shetland Islands
No records for South Havra [HU32], Papa, Oxna, Trondra and Hildasay [HU33], Vaila [HU24], Vementry [HU26], Papa Little [HU36], Lamba [HU38], Balta [HP60] and Muckle Flugga [HP61].

FAIR ISLE [HZ27]: *P. machaon*; *C. croceus*; *P. brassicae*; *P. rapae*; *P. icarus*[2]; *V. atalanta*; *C. cardui*; *A. urticae*; *I. io*; *M. jurtina*[5]; *C. pamphilus*[(1)]*.

MAINLAND SHETLAND [HU40]: *P. machaon*[2]; *C. croceus**; *P. brassicae*; *P. rapae**; *P. icarus*[2]; *V. atalanta*; *C. cardui*; *A. urticae*; *N. antiopa**; *I. io*; *C. tullia*[1]*(?); *D. plexippus**.

FOULA [HT93]: *P. brassicae*; *P. rapae*; *V. atalanta*; *C. cardui*; *N. antiopa*; *I. io*.

WEST BURRA [HU32]: *V. atalanta*.

BRESSAY [HU53]: *P. brassicae*; *P. napi*[4]; *V. atalanta*; *C. cardui*; *N. antiopa*; *I. io*.

NOSS [HU53]: *P. brassicae*; *V. atalanta*; *C. cardui*; *A. urticae*; *I. io*.

HALSAY [HU56]: *P. brassicae*; *V. atalanta*; *C. cardui*; *I. io*.

YELL [HU47]: *P. brassicae*; *P. rapae*; *V. atalanta*; *C. cardui*; *A. urticae*; *I. io*.

FETLAR [HU68]: *P. brassicae*; *V. atalanta*; *C. cardui*; *A. urticae*; *I. io*.

PAPA STOUR [HU15]: *P. brassicae*; *V. atalanta*; *C. cardui*.

MOUSA [HU42]: *V. atalanta*.

OUTER SKERRIES [HU67]: *P. brassicae*; *V. atalanta*; *A. urticae*; *I. io*.

UNST [HU95]: *P. brassicae*; *A. artaxerxes*[1]*(?); *V. atalanta*; *C. cardui*; *A. urticae*; *I. io*.

UYEA [HU69]: *P. brassicae*.

References
11, 17, 28, 29, 43, 46, 86–88, 96, 157, 199, 200, 205, 215, 222, 224, 225, 227, 228, 337, 346, 353, 354, 420, 427–431, 441, 442, 459, 477, 482, 483, 511, 514, 532, 540, 541, 544, 546, 549, 552, 553, 566, 567.

East Scotland and Firth of Forth
No records for Bell Rock [NO72] and Craigleith.

INCHKEITH [NT28]: *P. rapae*; *P. napi*[4]; *L. phlaeas*[(1)]; *P. icarus*[1]; *V. atalanta*; *C. cardui*; *A. urticae*.

INCHMICKERY [NT28]: *P. rapae*; *L. phlaeas*[(1)]; *V. atalanta*; *A. urticae*.

MAY [NT69]: *C. croceus*; *P. brassicae*; *P. rapae*; *P. napi*[4]; *L. phlaeas*[(1)]; *P. icarus*[1]; *V. atalanta*; *C. cardui*; *A. urticae*; *I. io*; *M. jurtina*[2]; *C. pamphilus*[1]*.

BASS ROCK [NT68]: *P. brassicae**; *P. rapae*; *P. napi*[4]; *L. phlaeas*[(1)]; *P. icarus*[1]; *V. atalanta*; *C. cardui*; *A. urticae**.

FIDRA [NT58]: *P. brassicae*; *P. rapae*; *P. napi*[4]; *V. atalanta*; *A. urticae*.

CRAMOND [NT17]: *C. croceus*; *P. brassicae*; *P. rapae*; *P. napi*[4]; *L. phlaeas*[(1)]; *P. icarus*[1]; *V. atalanta*; *C. cardui*; *A. urticae*; *I. io*; *M. jurtina*[2]; *C. pamphilus*[1].

INCHCOLM [NT18]: *P. brassicae*; *P. rapae*; *P. napi*[4]; *L. phlaeas*[(1)]; *P. icarus*[1]; *V. atalanta*; *C. cardui*; *A. urticae*; *I. io*; *M. jurtina*[2]; *C. pamphilus*[1].

References
22, 159, 160, 165, 167, 206, 452, 479, 511, 553.

Northumberland
No records from other Farne islands such as Megstones, Crumstones, Knivestone and Big Harcar [NU23], nor Coquet [NU20], St Mary's or Bait [NZ37].

LINDISFARNE [NU14]: *C. croceus*; *P. brassicae*; *P. rapae*; *P. napi*[4]; *A. cardamines*[2]; *L. phlaeas*[1]; *P. icarus*[1]; *V. atalanta*; *C. cardui*; *A. urticae*; *N. antiopa**?; *I. io*; *A. aglaja*[1]; *H. semele*[1]; *M. jurtina*[2]; *A. hyperantus*[1]; *C. pamphilus*[1].

INNER FARNE [NU23]: *P. brassicae*; *P. rapae*; *P. napi*[4]; *P. icarus*[1]*; *V. atalanta*; *C. cardui*; *A. urticae*; *I. io*; *A. aglaia*[1]; *L. megera*[(2)]; *H. semele*[1]; *M. jurtina*[2].

WEST WIDEOPENS [NU23]: *V. atalanta*; *A. urticae*.

LONGSTONE [NU23]: *V. atalanta*.

SOUTH WAMSES [NU23]: *A. cardamines*[2]; *A. urticae*.

BROWNSMAN [NU23]: *P. brassicae*; *P. rapae*; *P. napi*[4]; *P. icarus*[1]; *V. atalanta*; *C. cardui*; *A. urticae*; *I. io*; *H. semele*[1]; *M. jurtina*.

STAPLE [NU23]: *P. rapae*; *P. napi*[4]; *V. atalanta*; *C. cardui*; *A. urticae*; *I. io*.

NORTH WAMSES [NU23]: *V. atalanta*; *A. urticae*.

NORTHERN HARES [NU23]: *P. napi*[4]; *V. atalanta*.

References
91, 153, 424, 563.

C. Islands off Ireland

No records for Long [0092], Caher [0267], Inishkea South [0351], Inishkea North [0352], Gola [1472], Inishbofin [1483], Skerry Islands [2484], The Maidens [3441], Muck [3440], Copeland [3358], Mew [3368], Burial [3366], Guns [3353] and Irelands Eye [3224].

GREAT SALTEE [2099]: *L. sinapis*[2]; *C. croceus*; *P. brassicae*; *P. rapae*; *P. napi*[3]; *A. cardamines*[3]; *L. phlaeas*[2]; *C. minimus*; *P. icarus*[(2)]; *V. atalanta*; *C. cardui*; *A. urticae*; *N. antiopa**; *I. io*; *A. aglaja*[1]; *P. aegeria*[1]; *L. megera*[1]; *P. tithonus*[2]; *M. jurtina*[3]; *A. hyperantus*[1].

LESSER SALTEE [2099]: *P. brassicae*; *P. rapae*; *P. napi*[3]; *L. phlaeas*[2]; *P. icarus*[(2)]; *V. atalanta*; *C. cardui*; *A. urticae*; *M. jurtina*[3].

CLEAR [0092]: *C. hyale*; *C. croceus*; *G. rhamni*[2]; *P. brassicae*; *P. rapae*; *P. napi*[3]; *A. cardamines*[3]; *C. rubi*; *L. phlaeas*[2]; *C. minimus**; *P. icarus*[(2)]; *V. atalanta*; *C. cardui*; *A. urticae*; *I. io*; *A. aglaja*[1]; *E. aurinia*[2]; *P. aegeria*[1]; *L. megera*[1]; *H. semele*[(5)]; *P. tithonus*[2]; *M. jurtina*[3]; *A. hyperantus*[1]; *C. pamphilus*[1]; *D. plexippus*.

SHERKIN [1002]: *C. croceus*; *P. brassicae*; *P. rapae*; *P. napi*[3]; *A. cardamines*[3]; *C. rubi*; *L. phlaeas*[2]; *P. icarus*[(2)]; *C. argiolus*[2]; *V. atalanta*; *C. cardui*; *A. urticae*; *I. io*; *A. aglaja*[1]; *A. paphia*; *E. aurinia*[2]; *P. aegeria*[1]; *L. megera*[1]; *H. semele*[(5)]; *P. tithonus*[2]; *M. jurtina*[3]; *A. hyperantus*[1]; *C. pamphilus*[1].

WHIDDY [0095]: *P. brassicae*; *P. rapae*; *P. napi*[3]; *A. cardamines*[3]; *L. phlaeas*[2]; *P. icarus*[(2)]; *C. argiolus*[2]; *V. atalanta*; *A. urticae*; *I. io*; *P. aegeria*[1]; *M. jurtina*[3].

GARINISH [0095]: *P. brassicae*; *P. rapae*; *P. icarus*; *C. argiolus*[2]*; *C. cardui*.

DURSEY [0044]: *P. napi*[3]; *L. phlaeas*[2]*; *P. icarus*[(2)]*; *H. semele*[(5)]*.

BEAR [0074]: *C. rubi**; *A. urticae*; *I. io**; *A. aglaja*[1]; *H. semele*[(5)].

SCARRIF [0045]: *L. phlaeas*[2]; *P. icarus*[(2)]; *V. atalanta*; *C. cardui*; *A. urticae*; *L. megera*[1]; *M. jurtina*[3].

PUFFIN [0036]: *P. brassicae*; *P. rapae*; *A. urticae*; *I. io*; *M. jurtina*[3].

GREAT SKELLIG [0026]: *V. atalanta*.

VALENCIA [0047]: *C. croceus**; *P. rapae*; *P. napi*[3]; *A. cardamines*[3]; *C. rubi**; *L. phlaeas*[2]; *A. urticae*; *A. aglaja*[1]; *E. aurinia*[2]*; *P. aegeria*[1]; *L. megera*[1]; *H. semele*[(5)]*; *M. jurtina*[3].

GREAT BLASKET [0029]: *P. brassicae*; *P. rapae*; *P. napi*[3]; *L. phlaeas*[2]; *V. atalanta*; *C. cardui*; *A. urticae*; *I. io*; *P. aegeria*[1]; *M. jurtina*[3]; *A. hyperantus*; *C. pamphilus*[1].

INISHVICKILLANE [0029]: *P. brassicae*; *P. napi*[3]; *P. icarus*[(2)]; *V. atalanta*; *A. urticae*; *M. jurtina*[3].

MUTTON [0197]: *P. brassicae*; *P. napi*[3]; *M. jurtina*[3].

INISHMORE [0280]: *E. tages*[2]; *C. croceus*; *G. rhamni*[2]; *P. brassicae*; *P. rapae*; *P. napi*[3]; *A. cardamines*[3]; *L. phlaeas*[2]; *C. minimus*; *P. icarus*[(2)]; *V. atalanta*; *C. cardui*; *A. urticae*; *I. io*; *A. aglaja*[1]; *P. aegeria*[1]; *L. megera*[1]; *H. semele*[(4)]; *M. jurtina*[3]; *A. hyperantus*[1]; *C. pamphilus*[1].

INISHMAN [0290]: *P. brassicae*; *A. urticae*; *M. jurtina*[3].

INISHEER [0290]: *A. urticae**; *M. jurtina*[3]*.

GORUMNA [0282]: *P. brassicae*; *P. rapae*; *P. napi*[3]; *A. cardamines*[3]; *L. phlaeas*[2]; *P. icarus*[(2)]; *P. aegeria*[1]; *L. megera*[1]; *H. semele*[(5)]; *M. jurtina*[3]; *A. hyperantus*[1].

OMEY [0255]: C. cardui*; E. aurinia[2]*.

INISHARK [0246]: P. napi[3]; M. jurtina[3]; A. hyperantus; C. pamphilus[1].

INISHBOFIN [0256]: P. brassicae; P. rapae; P. napi[3]; L. phlaeas[2]; P. icarus[(2)]; C. cardui; A. urticae; P. aegeria[1]; L. megera; M. jurtina[3]; A. hyperantus; C. pamphilus[1].

INISHTURK [0257]: P. napi[3]; A. urticae; A. hyperantus; C. pamphilus[1].

CLARE [0268]: P. brassicae; P. napi[3]; L. phlaeas[2]; C. minimus; P. icarus[(2)]; C. cardui*; A. urticae; I. io; P. aegeria[1]*; L. megera[1]; H. semele[(5)]*; M. jurtina[3].

ACHILL [0350]: P. brassicae; P. rapae; P. napi[3]; C. rubi*; L. phlaeas[2]; C. minimus*; P. icarus[(2)]; C. argiolus[2]*; V. atalanta; C. cardui; A. urticae; I. io; L. megera[1]; M. jurtina[3]; A. hyperantus[1]; C. pamphilius[1]; C. tullia[1]*.

ARAN [1461]: P. brassicae; L. phlaeas[2]; V. atalanta; A. urticae; E. aurinia[2]; M. jurtina[3]; C. pamphilus[1].

CRUIT (near GOLA) [1472]: C. minimus; P. icarus[(2)]; A. urticae; I. io; A. aglaja[1]; E. aurinia[2]; H. semele[(5)]; M. jurtina[3]; C. pamphilus[1].

TORY [1484]: P. brassicae*; P. rapae*; P. napi[3]*; L. phlaeas[2]*; P. icarus[(2)]*; V. atalanta*; A. urticae*; I. io*; H. semele[(5)]*; M. jurtina[3]*.

INISHTRAHULL [2436]: P. brassicae; P. rapae; P. napi[3]; L. phlaeas[2]; V. atalanta; C. cardui; A. urticae; A. paphia; P. aegeria[1]; H. semele[(5)]); M. jurtina[3].

RATHLIN [3414]: C. croceus; P. brassicae; P. rapae; P. napi[3]; A. cardamines[3]; L. phlaeas[2]; P. icarus[(2)]; V. atalanta; C. cardui; A. urticae; A. aglaja[1]; P. aegeria[1]; L. megera[1]; H. semele[(5)]; M. jurtina[3]; A. hyperantus[1]; C. pamphilus[1].

JOHN'S (near COPELAND; also called CROSS or LIGHTHOUSE) [3358]: C. croceus; P. brassicae; P. rapae; P. napi[3]; A. cardamines[3]; V. atalanta; C. cardui; A. urticae; I. io; P. aegeria[1]; L. megera[1]; M. jurtina[3]; A. hyperantus[1].

LAMBAY [3235]: G. rhamni[2]*; P. brassicae; P. rapae; P. napi[3]; A. cardamines[3]; L. phlaeas[2]; P .icarus[(1)]; V. atalanta; C. cardui; A. urticae; I. io; A. aglaja[1]*; P. aegeria[1]; L. megera[1]; H. semele[(5)]*; M. jurtina[3]; A. hyperantus[1].

References
19, 20, 23–25, 89, 127, 152, 325, 330, 332, 333, 343, 344, 362–364, 445, 447.

3. SOURCE OF PERSONAL OBSERVATIONS

Date of letters and locations are given in brackets.

1 J. Asher, *Butterfly Net, Millennium Atlas Project* (3.v.96; Fair Isle, Shetland Islands, Orkney Islands and Isle of Man; *Nymphalis antiopa* and *Danaus plexippus* records for 1995).
2 M. Betts & S. Price (2.vi.95, 16.vi.95; Skokholm) including Macrolepidoptera lists in the Skokholm bird observatory reports of I. R. Downhill (1961), D. A. Scott (1968) A. Neale (1970) and J. Lawman (1977).
3 M. Bloomfield (13.v.94; Tresco).
4 W. W. Clynes (12.iii.96; Brownsman and Staple in Farnes).
5 P. N. Crow (1971; *C. minimus* near Douglas, Man).
6 T. C. Dunn (12.x.86; Lindisfarne).
7 J. A. Gibson (21.iv.95, 28.vi.95; Bute, Inchmarnock, Great Cumbrae, Little Cumbrae, Glas Eilean, Ailsa Craig, Horse, Lady, Holy, Pladda, Sanda, Davaar, Minard, Sgat Mor, Sheep, Glunimore, Burnt Islands).
8 R. D. & D. Graiff (29.ix.95; Colonsay and Oronsay).
9 P. Hackett (23.iii.94, 2.ix.94, 29.viii.95; Outer Hebrides).
10 P. T. Harding, ITE, BRC Data base at Monks Wood Experimental Station, Abbots Ripton, Huntingdon, to 16.iii.1988.
11 P. B. Hardy (24.v.95; Wight and Holy Island).
12 M. Hodges via S. McElwee (24.i.96; Inner Farne Islands).
13 D. C. Hulme (5.v.95; Handa, Tanera Mór, Tanera Beg, Eilean Fadadh Mór, Priest, Carn nan Sgeir, Mull, Staffa, Skye and Crowlin Isles).
14 M. Hull (18.iii.96; Herm).
15 T. A. Lavery & K. Cronin, Irish Lepidoptera Records Data base (15.iv.96; Lesser Saltee, Whiddy, Great Skellig, Great Blasket, Inishvickillane, Mutton, Inishark, Inishbofin, Inishturk).
16 R. M. Lockley *Letters from Skokholm* including reference to Dent (1947: 218) in M. Betts (16.vi.95; Skokholm).
17 R. Long and R. A. Austin (23.x.86; Channel Islands).
18 N. Mackenzie (16.vi.96; Garinish, Ireland).
19 M. McCormick via P. B. Hardy (6.ix.94; Bardsey).
20 S. McElwee compiling records from B. Baxter, N. Littlewood and S. Patterson (24.i.96; Outer Farne Islands).
21 M. Meakin (22.viii.95; Lundy).
22 J. R. Moon (16.v.94; Isles of Scilly; 22.viii.94; Alderney).
23 M. J. Morgan (19.x.88; Puffin Island).
24 R. Morris (16.xii.95; Inchcolm, Fidra, Inchkeith).
25 D. F. Owen (25.viii.95; Steepholm).
26 J. D. Parrack (1994; Farne Islands).
27 T. N. D. Peet (15.ii.87; Guernsey).
28 M. Pennington (ix.94; Shetland Islands).
29 A. J. Redfern (17.v.94; Isle of Wight).
30 I. Rippey (22.xii.88, 23.v.89, 28.vi.89, 7.vii.89, 18.viii.94; Irish islands including Rathlin, John's Copeland, Clare, Achill, Dursey, Cape Clear, Inishmore, Omey, Inistrahull, Tory, Cruit, Aranmore, Garinish and Lambay).
31 I. S. Robertson (28.x.80; Fair Isle. Copy from J. Heath).

32 R. I. Rutherford (26.ix.95; *P. napi* on Luing and Seil, *V. atalanta* on Eigg).
33 R. E. Saville (23.iii.96; Bass Rock and Cramond).
34 B. Shaw (15.ii.96; Hilbre from Hilbre Bird Observatory reports 1984–94; 19.ii.96; Isles of Scilly).
35 V. Seegar (20.iii.95; 9.vi.96; including records from Hilbre Bird Observatory); with A. Sawaik (9.vi.96; *C. rubi* on Mull).
36 I. Small and J. Taverner (15.vii.94; Hayling Island).
37 A. M. Smout, Fife Regional Council (1.xi.95; Inchcolm and Inchkeith).
38 A. Spalding (29.iii.95; St Michael's Mount).
39 G. Stringer and A. Wight with Farne Islands' wardens (2.ix.94, 10.ix.94, 27.x.94, 1.xi.94; Farne Islands).
40 G. Stringer from RSPB Vane Farm Nature Centre (8.xi.95; Fidra and Inchmickery in the Firth of Forth).
41 R. Sutcliffe (18.v.95; Ailsa Craig from G. Rodway and E.G. Hancock).
42 R. Sutton (6.v.94, 12.v.94; Isles of Scilly).
43 C. M. Tanner (14.viii.95; Lindisfarne and Longstone in Farnes).
44 W. G. Tremewan (24.viii.95; *P. icarus* on St Michael's Mount); with A. Spalding (24.vi.96; Looe Island).
45 G. Vicary (24.x.94; Channel islands).
46 J. Walton (19.v.95; Brownsman, Farnes).

Entry in proofs:
I. Bullock (20.xi.96; Grassholm, *C. cardui* on 25.vi.96).

BIBLIOGRAPHY

1 **Abercrombie, R. G.** 1953. *Callophrys rubi* in Iona. *Entomologist* **86**: 197.

2 **Ackery, P. R.** 1984. Systematic and faunistic studies on butterflies. pp. 9–21. *In* R. I. Vane-Wright & P. R. Ackery (Eds.), *The biology of butterflies*. London. Academic Press.

3 **Adkin, R.** 1892. [South London Ent. & NHS exhibition – Lepidoptera from Isles of Scilly]. *Entomologist*, **25**: 98.

4 —— 1894. On a collection of Lepidoptera from the Scilly Isles. *Entomologist* **27**: 9–12.

5 —— 1924. Divergence of character [*Aglais urticae* on Man]. *Entomologist's Rec. J. Var.* **37**: 11.

6 **Adkin, R. W.** 1897. *Colias edusa* in the Isles of Scilly. *Entomologist* **30**: 269.

7 —— 1911. *Colias edusa* and *Sphinx convolvuli* at Scilly. *Entomologist* **44**: 324.

8 **Agassiz, D.** 1982. *A revised list of the Lepidoptera of the Isles of Scilly*. Tresco, Isles of Scilly Museum Association.

9 **Allan, P. B. M.** 1949. Dispersal of *Lysandra coridon* Poda., (Lep., Lycaenidae). *Entomologist* **82**: 10.

10 **Angel, H.** 1975. *Photographing nature: insects*. London: Fountain Press.

11 **Anonymous.** 1993. Update [on Lepidoptera]. *Shetland Entomological Group Newsletter* No. 4 (Dec. 1993): 8–10.

12 **Askew, R. R.** 1974. Insects from Bardsey island. *Entomologist's Gaz.* **25**: 45–51.

13 **Bailey, N. T. J.** 1995. *Statistical methods in biology*. Cambridge: University Press.

14 **Baker, G.** 1900. *Lampides boeticus* bred in Guernsey. *Entomologist's mon. Mag.* **36**: 240.

15 **Baker, R. R.** 1969. The evolution of the migratory habit in butterflies. *J. anim. Ecol.*, **38**: 703–746.

16 —— 1978. *The evolutionary ecology of animal migration*. London: Hodder & Stoughton.

17 [**Baldwin, T.**] 1995. Shetland swallowtail. *Butterfly Conservation News* **60**: 16.

18 **Barber, K. E.** 1993. Peatlands as scientific archives of past biodiversity. *Biodiversity and Conservation* **2**: 474–489.

19 **Baring, C.** 1915. Notes on the fauna and flora of Lambay. *Ir. Nat.* **24**: 71.

20 —— 1918. Lepidoptera of Lambay. *Ir. Nat.* **27**: 63.

21 **Baxter, E. V. & Rintoul, L. J.** 1948. Peacock butterfly in Argyll and Mull. *Scott. Nat.* **1948**: 212.

22 —— 1950. Peacock and painted lady butterflies [on Isle of May]. *Scott. Nat.* **1950**: 50.

23 **Baynes, E. S. A.** 1957. The Lepidoptera of Tory island, county Donegal. *Entomologist* **90**: 310–317.

24 —— 1964. *A revised catalogue of Irish Macrolepidoptera*. Hampton, Middlesex: E. W. Classey.

25 —— 1973. Butterflies and moths. pp. 1–207. *In* J. T. R. Sharrock (Ed.), *The natural history of Cape Clear island*. Poyser, Berhampstead.

26 **Beavis, L. C.** 1976. Butterflies of the Isles of Scilly during August 1975. *Entomologist's Rec. J. Var.* **88**: 194–195.

27 **Beirne, B. P.** 1943a. The distribution and origin of the British Lepidoptera. *Proc. R. Ir. Acad.* **49** (B): 27–59.

28 —— 1943b. The relationship and origins of the Lepidoptera of the Outer Hebrides, Shetlands, Faroes and Iceland. *Proc. R. Ir. Acad.* **49** (B): 91–101.

29 —— 1945. The Lepidoptera of Shetland. *Entomologist's Rec. J. Var.* **57**: 37–40.

30 —— 1947. The origin and history of the British Macrolepidoptera. *Trans. R. ent. Soc.*, **98**: 275–372.

31 **Bennett, D. P. & Humphries, D. A.** 1974. *Introduction to field biology*. London: Edward Arnold.

32 **Bink, F. A.** 1992. *Ecologische Atlas van de dagvlinders van Noorwest-Europa*. Haarlem: Schuyt & Co.

33 **Bird, R. J.** 1980. Casual observations of insects in the Isle of Skye, July 1979. *Bull. amat. Ent. Soc.* **39**: 22–24.

34 **Birkett, N. L.** 1995. Is the Silver-studded Blue extinct in north-west England? *Butterfly Conserv. News* **59**: 24–25.

35 **Blackie, J. E. H.** 1951. Hebridean satyrids. *Entomologist* **84**: 264.

36 **Blackler, H.** 1940. Migration of large white butterflies observed at Hilbre Point, Hoylake, Cheshire. *Entomologist* **73**: 210.

37 **Blair, K. G.** 1925. Lepidoptera of the Scilly Isles. *Entomologist* **58**: 3–10, 38.

38 **Bland, K. P.** 1977. Lepidoptera from South Ronaldsay, Orkney. *Entomologist's Rec. J. Var.* **89**: 70–72.

39 **Blyth, S. F. P.** 1901. *Pieris daplidice* in Jersey. *Entomologist* **34**: 291.

40 **Bradley, J. D.** 1954. Lepidoptera records from Skokholm. *Entomologist* **87**: 146–147.

41 **Bradley, J. D. & Fletcher, D. S.** 1958. Lepidoptera records from the Isle of Rhum. *Entomologist* **91**: 126–131.

42 **Brakefield, P. M.** 1984. The ecological genetics of quantitative characters of *Maniola jurtina* and other butterflies. pp. 167–190. In R. I. Vane-Wright & P. R. Ackery (Eds.) *The Biology of butterflies*. London: Academic Press.

43 **Briggs, C. A.** 1884. A week's collecting on Unst. *Entomologist* **17**: 197–201.

44 **Bristow, C. R., Mitchell, S. H. & Bolton, D. E.** 1993. *Devon Butterflies* [includes Lundy]. Tiverton: Devon Books.

45 **Bristow, R.** 1994. *Satyrium w-album* (Knoch) (Lep.: Lycaenidae) in a moth trap. *Entomologist's Rec. J. Var.* **106**: 73.

46 **Bristowe, W. S.** 1931a. Notes on a migration of the 'Red Admiral' (*Pyrameis atalanta*). *Entomologist's mon. Mag.* **67**: 205–206.

47 —— 1931b. A small collection of insects from Lundy Island. *Entomologist's mon. Mag.* **67**: 232.

48 **Brock, J. P.** 1990. Origins and phylogeny of butterflies. pp. 209–233. *In* O. Kudrna (Ed.), *Butterflies of Europe*: **8**. *Introduction to Lepidopterology*. Wiesbaden: Aula-Verlag.

49 **Bryan, M. D.** 1987. Island interludes – Guernsey and the Isle of Mull in 1986. *Entomologist's Rec. J. Var.* **99**: 125–128.

50 **Buckstone, A. A. W.** 1938. Migration of *Polygonia c-album*. *Entomologist* **71**: 110.

51 **Campbell, J. L.** 1936. Immigrant Lepidoptera in the Outer Hebrides, 1936 (*Colias croceus* on Barra). *Entomologist* **69**: 265–266.

52 —— 1938. The Macrolepidoptera of the parish of Barra. *Scott. Nat.* **1938**: 153–163.

53 —— 1946. Catalogue of a collection of Macro-Lepidoptera made in the Hebrides between 1936 and the present date. *Entomologist* **79**: 49–53.

54 —— 1947a. Hebridean notes. *Entomologist* **80**: 116–117.

55 —— 1947b. Migrant Lepidoptera on the Isle of Canna during August 1947. *Entomologist* **80**: 269.

56 —— 1948a. Immigrant Lepidoptera in the Hebrides. *Scott. Nat.* **1948**: 47–48.

57 —— 1948b. On the spread of the peacock butterfly in the Highlands and Islands. *Scott. Nat.* **1948**: 126.

58 —— 1949a. Migrants in the Hebrides. *Entomologist* **82**: 189.

59 —— 1949b. Release of marked *Vanessa cardui*. *Entomologist* **82**: 209.

60 —— 1950. Lepidoptera taken at Heisker lighthouse in 1949. *Scott. Nat.* **1950**: 122.

61 —— 1951a. An experiment in marking migratory butterflies. *Entomologist* **84**: 1–6.

62 —— 1951b. *Callophrys rubi* on the Isle of Canna. *Entomologist* **84**: 159.

63 —— 1951c. *Pararge aegeria* on the Isle of Canna. *Entomologist* **84**: 237.

64 —— 1951d. *Vanessa atalanta* [Canna]. *Entomologist* **84**: 237.

65 —— 1952. Notes on Canna butterflies. *Entomologist* **85**: 263.

66 —— 1953. *Pieris brassicae* larvae in November and December; a further note. *Entomologist* **86**: 51.

67 —— 1954. The Macrolepidoptera of the Isle of Canna. *Scott. Nat.* **1954**: 101–121.
68 —— 1955a. *Vanessa cardui* on the Isle of Canna. *Entomologist* **88**: 231.
69 —— 1955b. *Vanessa atalanta* on the Isle of Canna. *Entomologist* **88**: 236.
70 —— 1958. Isle of Canna report for 1958. *Entomologist's Rec. J. Var.* **70**: 282–283.
71 —— 1967. Isle of Canna, 1966. *Entomologist's Rec. J. Var.* **79**: 97–99.
72 —— 1969a. Isle of Canna report for 1968. *Entomologist's Rec. J. Var.* **81**: 67–70.
73 —— 1969b. Rise and decline of *Vanessa io* in the Small Isles (Inner Hebrides). *Entomologist's Rec. J. Var.* **81**: 117–118.
74 —— 1969c. Isle of Canna report for 1969. *Entomologist's Rec. J. Var.* **81**: 323–325.
75 —— 1970. Macrolepidopera Cannae. The butterflies and moths of Canna. *Entomologist's Rec. J. Var.* **82**: 211–214, 235–242, 292–299.
76 —— 1972. Isle of Canna notes for 1970 and 1971. *Entomologist's Rec. J. Var.* **84**: 196–198.
77 —— 1975a. Isle of Canna report, 1972–1974. *Entomologists's Rec. J. Var.* **87**: 10–12.
78 —— 1975b. On the rumoured presence of the large blue butterfly (*Maculinea arion*) in the Hebrides. *Entomologist's Rec. J. Var.* **87**: 161–166.
79 —— 1978. *Pieris rapae* on the Isle of Canna. *Entomologist's Rec. J. Var.* **89**: 279.
80 —— 1980. Isle of Canna: invasion of painted ladies and red admirals. *Entomologist's Rec. J. Var.* **92**: 256–257.
81 —— 1984. Isle of Canna, 1984: the peacock returns. *Entomologist's Rec. J. Var.* **96**: 187.
82 **Campbell, R. C.** 1989. *Statistics for biologists.* Cambridge: University Press.
83 **Campbell-Taylor, J. E.** 1931a. *Melitaea aurinia* Rott. *Entomologist* **64**: 188–189.
84 —— 1931b. The flying power of *Melitaea aurinia. Entomologist* **64**: 278.
85 **Carpenter, G. D. H.** 1943. New records of insects and woodlice from Lundy Island. *Entomologist's mon. Mag.* **87**: 121–123.
86 **Carrington, J. T.** 1880. The Lepidoptera of the Shetland Isles. *Entomologist* **13**: 238.
87 —— 1881. [Lepidoptera of Shetland Isles and Hebrides.] *Entomologist* **14**: 303.
88 —— 1883. Lepidoptera on Unst. *Entomologist* **16**: 236–237.
89 **Carter, D. J.** 1967. The Lepidoptera, Trichoptera and Odonata of Cape Clear Island. *Cape Clear Bird Obs. Rep.* **8**: 45–46.
90 **Carter, H. H., Smith, V. C. & Spence, T. F.** 1960. Puffin Island 1959. *Proc. Bgham nat. Hist. phil. Soc.* **19**: 37–53.
91 **Cartwright, R. A. & Cartwright, D. B.** 1976. *The Holy Island of Lindisfarne and the Farne Islands.* Newton Abbot: David & Charles.
92 **Castle-Russell, S. G.** 1932. Flying habits of *Melitaea aurinia. Entomologist* **65**: 42–43.
93 **Chalmers-Hunt, J. M.** 1970. The butterflies and moths of the Isle of Man. *Trans. Soc. Br. Ent.* **19**: 1–171.
94 **Cheesman, E. M.** 1898. Lepidoptera captures in the Orkney Isles. *Entomologist's Rec. J. Var.* **10**: 204–206.
95 **Clapham, A. R., Tutin, T. G. & Moore, D. M.** 1989. *Flora of the British Isles.* Cambridge; Cambridge University Press.
96 **Clarke, W. E.** 1897. *Vanessa antiopa* in Shetland. *Ann. Scot. nat. Hist.* **1897**: 48.
97 **Clutterbuck, C. G.** 1940. Lepidoptera in the Isles of Scilly in August 1939. *Entomologist* **73**: 129–131.
98 **Colvin, M. & Reavey, D.** 1993. *A directory for entomologists.* Middlesex: The Amateur Entomologists' Society. Pamphlet No. 14.
99 **Conder, P. J.** 1949. Observations on a migration of *Pieris brassicae* L. at Skokholm Island, Pembrokeshire, in August 1947. *Proc. R. ent. Soc. Lond.* (A) **24**: 35–38.
100 **Coney, G. B.** 1904. Lepidoptera in Jersey, 1903. *Entomologist* **37**: 127–131.
101 **Conservation Committee for Butterfly Conservation.** 1995. Lepidoptera restoration: Butterfly Conservation's policy, code of practice and guidelines for action. *Butterfly Conservation News* **60**: 20–21.
102 **Cooter, J.** 1978. Some insects from Ailsa Craig. *Entomologist's mon. Mag.* **113**: 168.
103 **Corbet, A. S.** 1945. The Lepidoptera of St Kilda. *Entomologist* **78**: 166–168.
104 **Courtney, S. P.** 1980. Studies on the biology of the butterflies *Anthocharis cardamines* L and *Pieris napi* L. in relation to speciation in the Pierinae. Ph.D. thesis, University of Durham.

105 **Courtney, S. P., Hill, C. P. & Westerman, A.** 1982. Pollen carried for long periods by butterflies. *Oikos* **38**: 260–263.
106 **Cowin, W. S.** 1955. Our missing butterflies [Man]. *Peregrine* **2**: 6–8.
107 **Craggs, J. D.** 1982. *Hilbre the Cheshire island, its history and natural history*. Liverpool, Liverpool University Press.
108 **Crewe, H. H.** 1877a. Entomology at Tresco and the Scilly Isles. *Entomologist* **10**: 295–297.
109 —— 1877b. Notes on the Lepidoptera of the Scilly Isles. *Entomologist's mon. Mag.* **14**: 148–150.
110 **Curtis, W. P.** 1931. The flying powers of *Melitaea aurinia*. *Entomologist* **64**: 211–212.
111 —— 1932. The flying powers of *Melitaea aurinia*. *Entomologist* **65**: 73–80.
112 **Dale, C. W.** 1884a. Capture of insects in the Hebrides and in St Kilda. *Scott. Nat.* **1884**: 284.
113 —— 1884b. Captures on South Uist and St Kilda. *Entomologist's mon. Mag.* **20**: 213–214.
114 —— 1884c. Additions to the entomology of the Isle of Harris. *Entomologist's mon. Mag.* **20**: 256.
115 —— 1884d. Captures in the Isle of Skye. *Entomologist's mon. Mag.* **20**: 257.
116 —— 1889. The insect fauna of St Kilda. *Entomologist* **22**: 12–13.
117 —— 1890. Insects in the Scilly Isles. *Entomologist's mon. Mag.* **26**: 238.
118 **Dannreuther, T.** 1933. Migration records, 1932 (*Pieris rapae* on Round Island). *Entomologist* **66**: 268.
119 —— 1935. Migrations records, 1934 (*Cynthia cardui* on Sandray). *Entomologist* **68**: 209.
120 —— 1936. Migration records, 1936. *Entomologist* **69**: 225–230.
121 —— 1939. Migration record for 1938 [*Aglais urticae* on North Rona]. *Entomologist* **72**: 119–122.
122 —— 1939. Migration records, 1939. *Entomologist* **72**: 273–283.
123 —— 1944. Migration records, 1943 [*Colias croceus* on Round Island]. *Entomologist* **77**: 112.
124 —— 1948a. Migration records, 1947 [*Cynthia cardui* on Chicken Rock, Man]. *Entomologist* **81**: 73–83.
125 —— 1948b. Insect migration: Scotland an end point. *Scott. Nat.* **1948**: 74–81.
126 —— 1951. Insect migration in 1950. *Scott. Nat.* **1951**: 122–127.
127 —— 1950. Migration records, 1950 [Great Saltee]. *Entomologist* **84**: 85–90.
128 **Darlington, A.** 1954. Ecological notes on some insects of Bardsey during August 1954. *Rep. Bardsey Bird Fld Obs.* **2**: 35–38.
129 **Dempster, J. P., Lakhani, K. H. & Coward, P. A.** 1986. The use of chemical composition as a population marker in insects: a study of the brimstone butterfly. *Ecol. Entomol.* **11**: 51–56.
130 **Demuth, R. P.** 1974. Late autumn in the Scilly Isles. *Entomologist's Rec. J. Var.* **86**: 72–73.
131 **Dennis, R. L. H.** 1972. *Plebejus argus* (L.) *caernensis* Thompson. A stenoecious geotype. *Entomologist's Rec. J. Var.* **84**: 132–140.
132 —— 1974. *Lysandra coridon* (Poda) and *L. bellargus* (Rott.) in North Wales? *Entomologist's Rec. J. Var.* **86**: 24.
133 —— 1977. *The British butterflies. Their origin and establishment*. Oxon, Faringdon: E. W. Classey.
134 —— 1982a. Patrolling behaviour in orange tip butterflies within the Bollin valley in north Cheshire, and a comparison with other pierids. *Vasculum* **67**: 17–25.
135 —— 1982b. Observations on habitats and dispersal made from oviposition markers in north Cheshire *Anthocharis cardamines* (L.) (Lep., Pieridae). *Entomologist's Gaz.* **33**: 151–159
136 —— 1985. Small plants attract attention! Choice of egg-laying sites in the green-veined white butterfly (*Artogeia napi*) (L.) (Lep: Pieridae). *Bull. amat. Ent. Soc.* **44**: 77–82.
137 —— 1986. Motorways and cross-movements. An insect's 'mental map' of the M56 in Cheshire. *Bull. amat. Ent. Soc.* **45**: 228–243.
138 —— (Ed.) 1992. *The ecology of butterflies in Britain*. Oxford: Oxford University Press.
139 —— 1993. *Butterflies and climate change*. Manchester: Manchester University Press.
140 **Dennis, R. L. H. & Bardell, P.** 1996. The impact of extreme weather events on local populations of *Hipparchia semele* (L.) (Nymphalidae) and *Plebejus argus* (L.) (Lycaenidae): hindsight, inference and lost opportunities. *Entomologist's Gaz.* **47**: 211–225.
141 **Dennis, R. L. H. & Bramley, M. J.** 1985. The influence of man and climate on dispersion

patterns within a population of adult *Lasiommata megera* (L.) (Satyridae) at Brereton Heath, Cheshire (UK). *Nota lepid.* **8**: 309–324.

142 **Dennis, R. L. H. & Shreeve, T. G.** 1989. Butterfly morphology variation in the British Isles: the influence of climate, behavioural posture and the hostplant-habitat. *Biol. J. Linn. Soc.* **38**: 323–348.

143 —— 1991. Climate change and the British butterfly fauna: opportunities and constraints. *Biol. Conserv.* **55**: 1–16.

144 **Dennis, R. L. H. & Williams, W. R.** 1986. Butterfly 'diversity': regressing and a little latitude. *Antenna* **10**: 108–112.

145 **Dennis, R. L. H., Williams, W. R. & Shreeve, T. G.** 1991. A multivariate approach to the determination of faunal structures among European butterfly species (Lepidoptera, Rhopalocera). *Zool. J. Linn. Soc.* **101**: 1–49.

146 **Dickson, R.** 1992. A Lepidopterist's handbook. *Amat. Ent.* **13**.

147 **Dobson, K. S.** 1981. *Pararge aegeria insula* Howarth (Lep., Satyridae) in 1980. *Entomologist's Gaz.* **32**: 113.

148 **Dowdeswell, W. H.** 1936. Lepidoptera of Cara Island. *Entomologist* **69**: 49–53.

149 —— 1981. *The Life of the Meadow Brown.* London: Heinemann.

150 **Dowdeswell, W. H. & Ford E. B.** 1948. Butterfly migration noted in the Isles of Scilly in 1947. *Entomologist* **81**: 141.

151 **Dunn, T. C.** 1965. Lepidoptera seen on the Isle of Colonsay 1964. *Entomologist* **98**: 243–247.

152 —— 1969. Collecting in the Aran Islands. *Entomologist's Gaz.* **20**: 271–278.

153 **Dunn, T. C. & Parrack, J. D.** 1986. The moths and butterflies of Northumberland and Durham. Part I: Macrolepidoptera. *Vasculum.* Suppl. **2**.

154 **Dymond, J. N.** 1974. Butterflies in 1972. *Rep. Lundy Fld Soc.* **23**: 38.

155 —— 1975. Butterflies and moths. *Rep. Lundy Fld Soc.* **24**: 24–27.

156 **Eales, H. T.** 1995. A revision of the status of the large heath butterfly (*Coenonympha tullia*) in Northumberland. Unpublished report.

157 **[Editors]** 1888. The Lepidoptera of the Outer Hebrides, Orkney and Shetland. *Scott. Nat.* **1888**: 298–304.

158 —— 1919. The butterflies of the Isle of Coll, Inner Hebrides. *Scott. Nat.* **1919**: 64.

159 **Eggeling, W. J.** 1957. A list of the butterflies and moths recorded from the Isle of May. *Scott. Nat.* **1957**: 75–83.

160 —— 1985. *The Isle of May. A Scottish nature reserve.* Perthshire, Larien Press.

161 **Ellis, H. A.** 1994. The status of the wall brown butterfly, *Lasiommata megera*, in Northumberland, 1965–91, in relation to local weather. *Trans. nat. Hist. Soc. Northumb.* **56**: 135–152.

162 **Ellis, P. M.** 1896. *Vanessa antiopa* at Skye. *Entomologist* **29**: 316.

163 **Ellis, R. G.** 1981. Increase of peacock butterfly on island of Arran. *W. Nat.* **1981**: 27.

164 **Emmet, A. M. & Heath, J.** (Eds) 1989. *The moths and butterflies of Great Britain and Ireland.* **7** (1) Colchester, Essex: Harley Books.

165 **Evans, W.** 1912a. The painted lady butterfly at the Isle of May. *Scott. Nat.* **1912**: 261–262.

166 —— 1912b. Some Lepidoptera and other insects from St Kilda. *Scott. Nat.* **1912**: 262.

167 —— 1918. Insects and other terrestrial invertebrates of the Bass Rock. *Scott. Nat.* **1918**: 259–265.

168 —— 1922. Breeding of the painted lady in Arran. *Scott. Nat.* **1922**: 27.

169 **Evans, W. & Grimshaw, P. H.** 1916. Notes on insects captured in the Island of Raasay. *Scott. Nat.* **1916**: 299–300.

170 **Fassnidge, W.** 1933. Immigrant Lepidoptera [Round Island]. *Entomologist* **66**: 19–20.

171 **Fearnehough, T. D.** 1937. Single brooded *Polyommatus icarus*; Sheffield. *Entomologist's Rec. J. Var.* **49**: 83–84.

172 —— 1972. Butterflies in the Isle of Wight. *Entomologist's Rec. J. Var.* **84**: 57–64, 102–109.

173 **Fenton, E. W.** 1948. Miscellaneous zoological notes. *Scott. Nat.* **1948**: 223–225.

174 **Fitter, R., Fitter, A. & Blamey, M.** 1978. *The wild flowers of Britain and northern Europe.* London: Collins.

175 **Fitter, R., Fitter, A. & Farrer, A.** 1984. *Grasses, sedges, rushes and ferns of Britain and Northern Europe*. London: Collins.
176 **Fletcher, T. B.** 1901. The naval manoeuvres of 1900, from an entomological point of view [Isle of Sheppey]. *Entomologist* **34**: 71–73.
177 **Foggit, G. T.** 1981. Danaus plexippus in the Scilly Isles in 1981. *Entomologist's Rec. J. Var.* **93**: 202.
178 **Forbes, W. A.** 1876. Notes on the entomology of Skye. *Scott. Nat.* **1876**: 262–264.
179 **Ford, E. B.** 1945. *Butterflies*. London: Collins. (Third edition, 1957).
180 —— 1964. *Ecological genetics*. London: Methuen.
181 **Forrest, J. E., Waterston, A. R. & Watson, E. V.** (Eds) 1936. The natural history of the Isle of Barra, Outer Hebrides. *Proc. R. phys. Soc. Edinb.* **22**: 241–296.
182 **Frazer, J. F. D.** 1939. *Nymphalis io* at night. *Entomologist* **72**: 20.
183 **Freeman, R. B.** 1976. The obvious insects in Alderney. (1) Butterflies. *Q. Bull. Alderney Soc.* **11**: 7–9.
184 **Fremlin, H. S.** 1900. Collecting in the Isle of Lewis. *Entomologist* **33**: 37–39.
185 **French, R. A.** 1958. Migration records. *Entomologist* **91**: 101–109.
186 **Frohawk, F. W.** 1935. *Nymphalis polychloros* in the Scilly Isles: is it a migrant? *Entomologist* 68: 262.
187 **Frost, M. P. & Madge, S. C.** 1991. *Butterflies of south-east Cornwall* [includes Looe island]. The Caradon Field & Natural History Club.
188 **Fryer, J. C. F.** 1926. Lepidoptera in the Scilly Isles. *Entomologist* **59**: 82.
189 **Garland, S. P.** 1981. *Butterflies of the Sheffield area*. Sheffield: Sorby Natural History Society and Sheffield Museums.
190 **Garrad, L. S.** 1973. Notes on the natural history of Raasay and Scalpay. *Glasg. Nat.* **19**: 7–11.
191 **Gibson, J. A.** 1952. The butterflies of Ailsa Craig. *Scott. Nat.* **1952**: 112–113.
192 —— 1976. Grayling butterflies on Ailsa Craig. *W. Nat.* **1976**: 114.
193 —— 1982a. Notes on the butterflies of the Cumbrae islands. *W. Nat.* **1982**: 5–8.
194 —— 1982b. Butterfly notes from Horse island, Ayrshire. *W. Nat.* **1982**: 9.
195 —— 1982c. Butterfly notes from the Sanda island group, Kintyre. *W. Nat.* **1982**: 10.
196 —— 1982d. Butterflies flying over the sea towards Clyde islands. *W. Nat.* **1982**: 11–12.
197 —— 1990. The butterflies of the island of Bute. *Trans. Butesh. nat. Hist. Soc.* **23**: 41–45.
198 —— 1992. Clouded yellow butterfly at Kilchattan bay; new record for the island of Bute. *Scott. Nat.* **1992**: 127.
199 **Goater, B.** 1969. Entomological excursions to the Shetlands, 1966 and 1968. *Entomologist's Gaz.* **20**: 73–82.
200 —— 1973. Some further observations on Shetland Lepidoptera, 1972. *Entomologist's Gaz.* **24**: 7–12.
201 —— 1974. A remarkable year [Orkney]. *Entomologist's Rec. J. Var.* **86**: 214–219, 234–239.
202 **Graves, P .P.** 1930. The British and Irish *Maniola jurtina* L. (Lep., Satyridae). *Entomologist* **68**: 49–54, 75–81.
203 **Gregson, C. S.** 1885. Notes on the Lepidoptera taken by E. R. Curzon Esq. at the island of Hoy, one of the Orkney islands, during the summer of 1885. *Young Nat.* **6**: 272–278.
204 **Griffiths, G. C. & Bartlett, C.** 1914. The natural history of Steepholm entomology. *Proc. Bristol Nat. Soc.* **4**: 143–149.
205 **Grimshaw, P. H.** 1906. Insects from Fair Isle. *Ann. Scot. Nat. Hist.* **1906**: 118.
206 —— 1908. A contribution to the insect fauna of the Isle of May. *Ann. Scot. Nat. Hist.* **1908**: 89–90.
207 —— 1920. Notes on the insect fauna of South Uist. *Scott. Nat.* **1920**: 85–89.
208 —— 1933. Scottish insect immigration records. *Scott. Nat.* **1933**: 173–181.
209 **Grimshaw, P. H. & Evans, W.** 1916. Notes on insects captured in the Isle of Raasay. *Scott. Nat.* **1916**: 299–300.
210 **Grimwood, K. W.** 1965. Fritillary in garden. *Entomologist's Rec. J. Var.* **77**: 240.
211 **Haggart, G. D.** 1933. Notes on Arran Lepidoptera. *Entomologist* **66**: 28–30.
212 **Hale, J. W. & Hicks, M. E.** 1995. Fall of the painted lady. *Entomologist's Rec. J. Var.* **107**: 188–191.

213 **Hall, D. W.** 1950. A note on the distribution of the peacock butterfly *Nymphalis io* in Scotland. *Scott. Nat.* **1950**: 109–110.
214 **Hallet, H. M.** 1930. A contribution to the entomological fauna of Skomer island. *Trans. Cardiff Nat. Soc.* **61** (1928): 72–76.
215 **Hanbury, F. J. H.** 1895. Notes on the Lepidoptera observed during a short botanical tour in west Sutherland, the Orkneys and Shetlands. *Entomologist's mon. Mag.* **31**: 1–12.
216 **Hancock, E. G.** 1978. Additional insects recorded for Puffin island. *Proc. Bgham nat. Hist. Soc.* **23**: 239–241.
217 —— 1995. *Terrestrial invertebrates from Ailsa Craig.* Glasgow Museum & Art Gallery (unpublished list).
218 **Hancock, G. L. R.** 1923. Contributions towards a list of the insect fauna of the South Ebudes. The Lepidoptera. *Scott. Nat.* **1923**: 125–132.
219 **Hanski, I.** 1994. Patch-occupancy dynamics in fragmented landscapes. *Trends Ecol. Evol.* **9**: 131–135.
220 **Hanski, I., Kuusaari, M. & Nieminen, M.** 1994. Metapopulation structure and migration in the butterfly *Melitaea cinxia*. *Ecology* **75**: 747–762.
221 **Hanski, I. & Thomas, C. D.** 1994. Metapopulation dynamics and conservation: a spatially explicit model applied to butterflies. *Biol. Conserv.* **68**: 167–180.
222 **Hardy, D. E.** 1956. The Lepidoptera of Fair Isle. *Entomologist* **89**: 261–269.
223 **Hardy, P. B., Hind, S. H. & Dennis, R. L. H.** 1993. Range extension and distribution-infilling among selected butterfly species in north-west England: evidence for inter-habitat movements. *Entomologist's Gaz.* **44**: 247–255.
224 **Hare, E. J.** 1963. Shetland Macro-Lepidoptera. *Entomologist's Rec. J. Var.* **75**: 238.
225 **Harper, M. W.** 1974. *Polyommatus icarus* in Shetland. *Entomologist's Rec. J. Var.* **86**: 120.
226 **Harrison, S.** 1991. Local extinction in a metapopulation context: an empirical evaluation. *Biol. J. Linn. Soc.* **42**: 73–88.
227 **Harvey, P.** 1990. Butterflies on Fair Isle 1978-1989. *Rep. Fair Isle Bird Obs.* **42**: 55–57.
228 **Harvey, P., Riddiford, N. & Riddiford, E.** 1992. Butterflies on Fair Isle 1990–92. *Rep. Fair Isle Bird Obs.* **44**: 73–74.
229 **Heath, J., Pollard, E. & Thomas, J. A.** 1984. *Atlas of butterflies in Britain and Ireland.* London: Viking.
230 **Heath, P. M.** 1975. A visit to Skomer. *Entomologist's Rec. J. Var.* **87**: 67–70.
231 —— 1976. Skomer Lepidoptera. *Nature Wales* **15**: 8–14.
232 **Heckford, R. J.** 1978. Scilly records for 1977. *Entomologist's Rec. J. Var.* **90**: 157.
233 **Hedges, A. V.** 1947. A list of Manx Lepidoptera. *Entomologist* **80**: 44–46, 62–66, 89–94.
234 **Hedges, J.** 1980. Notes on the Lepidoptera in the Isle of Man in 1980. *Entomologist's Rec. J. Var.* **92**: 303.
235 —— 1981. *Danaus plexippus* in the Isle of Man. *Entomologist's Rec. J. Var.* **93**: 202.
236 **Herbert, C.** 1993. *Butterflies of the London Borough of Barnet: A provisional atlas.* East Barnet, Herts: Herts & Middlesex Wildlife Trust.
237 **Heron, A. M. R.** 1956. Lepidoptera noted near Tenby, Pembrokeshire [Grassholm]. *Nature Wales* **2**: 350–351.
238 **Heslop Harrison, J.** 1947. Butterflies on the Isle of Tiree. *Entomologist's Rec. J. Var.* **59**: 140.
239 **Heslop Harrison, J. W.** 1937a. Rhopalocera on the Isle of Scalpay, with an account of the occurrence of *Nymphalis io* L. on Raasay. *Entomologist* **70**: 1–4.
240 —— 1937b. A contribution to our knowledge of the Lepidoptera of the islands of Eigg, Canna and Sanday. *Entomologist's Rec. J. Var.* **49**: 29–31.
241 —— 1937c. The natural history of the Isle of Raasay and of the adjacent islands of South Rona, Scalpay, Fladday and Longay. *Proc. Univ. Durham phil. Soc.* **9**: 246–251.
242 —— 1938a. The Rhopalocera of the islands of Coll, Canna, Sanday, Rhum, Eigg, Soay and Pabbay (Inner Hebrides) and of Barra, Mingulay and Berneray (Outer Hebrides). *Entomologist* **71**: 18–20.
243 —— 1938b. *Argynnis selene*, a butterfly new to the Isle of Rhum. *Entomologist* **71**: 213.
244 —— 1938c. The Rhopalocera of the islands of Rhum, Eigg, Muck, Eilean nan Each and

Heiskeir (Inner Hebrides) and of Harris, North Uist, South Uist, Eriskay, Taransay and the Monach islands (Outer Hebrides). *Entomologist* **71**: 265–267.
245 —— 1939. *Argynnis aglaia* and *Pieris napi* flying out to sea. *Entomologist* **72**: 59.
246 —— 1940a. Lepidoptera of the Isle of Handa. *Entomologist* **73**: 44–45.
247 —— 1940b. The common blue in the Isles of Pabbay, Fladday and Fiaray (Barra Isles). *Entomologist* **73**: 50.
248 —— 1940c. *Pieris napi* flying in dusk on the Isle of Coll. *Entomologist* **73**: 91.
249 —— 1940d. More Hebridean days. 1: The Isle of Muldoanich and the Uidh peninsula of Vatersay [includes Uinessan]. *Entomologist* **73**: 101–103.
250 —— 1940e. *Aglais urticae* much earlier on the Isle of Coll than on the Isle of Tiree. *Entomologist* **73**: 115.
251 —— 1940f. *Eumenis semele* L. and *Maniola jurtina* L. on the Isles of Coll and Gunna, Inner Hebrides. *Entomologist's Rec. J. Var.* **52**: 20.
252 —— 1940g. *H. phlaeas* in the Isle of Colonsay. *Entomologist's Rec. J. Var.* **52**: 129.
253 —— 1940h The range of *Argynnis aglaia* in the Outer Hebrides. *Entomologist's mon. Mag.* **76**: 274.
254 —— 1941a. More Hebridean days. II. The Isle of Benbecula. *Entomologist* **74**: 1–5.
255 —— 1941b. Immigrant Lepidoptera in the Inner and Outer Hebrides in 1940. *Entomologist* **74**: 19.
256 —— 1941c. More Hebridean days. III. The Isles of Tiree and Gunna. *Entomologist* **74**: 97–100.
257 —— 1941d. Lepidoptera in the Isle of Islay. *Entomologist* **74**: 238.
258 —— 1941e. *Colias croceus* F. in the Isle of Harris. *Entomologist* **74**: 278.
259 —— 1941f. The distribution and habits of *Callophrys rubi* L. in the Isle of Rhum. *Entomologist's Rec. J. Var.* **53**: 86–88.
260 —— 1942a. *Vanessa cardui* in the Orkney Isles. *Entomologist* **75**: 16.
261 —— 1942b. The irregularity of broods of *Pieris brassicae* L. on the Isles of Lewis, Harris and Scarp in 1941. *Entomologist* **75**: 33.
262 —— 1942c. Lepidoptera on the Isle of Scotasay. *Entomologist* **75**: 36.
263 —— 1942d. Butterflies in the Isles of Rhum and South Uist during May 1942. *Entomologist* **75**: 185.
264 —— 1942e. *Argynnis aglaia* race *scotica* Watkins in the western isles of Scotland. *Entomologist's Rec. J. Var.* **54**: 44–45.
265 —— 1942f. Variation in Outer Hebridean larvae of *Aglais urticae* L. and the fate of Rhum pupae of *Pieris napi* L. reared in 1941. *Entomologist's Rec. J. Var.* **54**: 107.
266 —— 1942g. The range of *Argynnis selene* in the Isle of Rhum. *Entomologist's mon. Mag.* **78**: 280.
267 —— 1943a. Lepidoptera on the Isle of Baleshare. *Entomologist* **76**: 58.
268 —— 1943b. August captures of *Coenonympha tullia* Müller var. *scotica* Stdgr. in the Outer Hebrides. *Entomologist* **76**: 106.
269 —— 1943c. Other Lepidoptera noted in North Uist in 1942. *Entomologist's Rec. J. Var.* **55**: 26.
270 —— 1943d. The range of *Euphydryas aurinia* (Lep: Nymphalidae) in the Hebrides and some possible deductions therefrom. *Entomologist's Rec. J. Var.* **55**: 27.
271 —— 1943e. Immigrant Lepidoptera in the Inner and Outer Hebrides. *Entomologist's Rec. J. Var.* **55**: 108.
272 —— 1944. An apparently new foodplant for *Vanessa cardui*. *Entomologist's Rec. J. Var.* **56**: 25.
273 —— 1945a. Remarks on certain Lepidoptera from the western isles of Scotland. *Entomologist* **78**: 18–21.
274 —— 1945b. The peacock butterfly (*Nymphalis io* L.) reaches the Isle of Rhum. *Entomologist* **78**: 46.
275 —— 1945c. Lepidoptera new to, or rare in, the Outer Hebrides. *Entomologist's Rec. J. Var.* **57**: 1–3.
276 —— 1945d. The foodplants of *Coenonympha tullia* Müller. *Entomologist's Rec. J. Var.* **57**: 125–126.
277 ——1946a. *Maniola jurtina* goes for a sail. *Entomologist* **79**: 69–70.
278 —— 1946b. The drinking of *Pieris napi* and the abundance of *Pieris brassicae* larvae on Rhum. *Entomologist* **79**: 70.

279 —— 1946c. Noteworthy Lepidoptera from the Isle of Rhum, with some notes on the insects captured on the adjacent islands. *Entomologist* **79**: 147–151.
280 —— 1946d. The geographical distribution of certain Hebridean insects and deductions to be made from it. *Entomologist's Rec. J. Var.* **58**: 18–21.
281 —— 1946e. The Lepidoptera of the Isles of Coll, Tiree and Gunna, with some remarks on the biogeography of the islands. *Entomologist's Rec. J. Var.* **58**: 57–61.
282 —— 1946f. Immigrant Lepidoptera in the Inner and Outer Hebrides in 1946. *Entomologist's Rec. J. Var.* **58**: 143.
283 —— 1946g. The range of *Pieris napi* in the Isle of South Uist. *Entomologist's Rec. J. Var.* **58**: 144.
284 —— 1947a. Early spring insects from the Isle of Rhum, with some remarks on the woodland fauna of the island. *Entomologist* **80**: 1–4.
285 —— 1947b. The status of the peacock butterfly in the small isles parish of Inverness-shire. *Entomologist* **80**: 13.
286 —— 1947c. *Colias hyale* L. and *C. croceus* Fourc. on the Isle of Coll, Inner Hebrides. *Entomologist's Rec. J. Var.* **59**: 139.
287 —— 1947d. *Pieris napi* L. in the Inner and Outer Hebrides. *Entomologist's Rec. J. Var.* **59**: 139.
288 —— 1947e. *Polyommatus icarus* race *clara* Tutt on the Isle of Ronay, Outer Hebrides. *Entomologist's Rec. J. Var.* **59**: 140.
289 —— 1947f. The Pleistocene races of certain British insects and distributional overlapping. *Entomologist's Rec. J. Var.* **59**: 141–145.
290 —— 1948a. Lepidoptera in the Inner and Outer Hebrides during the year 1947. *Entomologist* **81**: 1–6.
291 —— 1948b. Inbreeeding in Hebridean Lepidoptera populations. *Entomologist's Rec. J. Var.* **60**: 46–50.
292 —— 1948c. A new race of *Coenonympha pamphilus* for the Hebrides. *Entomologist's Rec. J. Var.* **60**: 111–112.
293 —— 1949a. A contribution to our knowledge of the Lepidoptera of the Isles of Lewis and Harris. *Entomologist* **82**: 16–20.
294 —— 1949b. Further observations on the Lepidoptera of the Scottish Western Isles. *Entomologist* **82**: 265–268.
295 —— 1949c. The foodplants of *Callophrys rubi* in the Inner Hebrides. *Entomologist's Rec. J. Var.* **61**: 6.
296 —— 1949d. Rhopalocera in the western Scottish Isles in 1948, with an account of two new forms of *Pararge aegeria* L. (Lep., Satyridae). *Entomologist's mon. Mag.* **85**: 25–28.
297 —— 1950a. Immigrant Lepidoptera in the Scottish Western Isles in 1949. *Entomologist* **83**: 70–71.
298 —— 1950b. Lepidoptera in the Outer Hebrides. *Entomologist* **83**: 241–245.
299 —— 1950c. Observations on the ranges, habitats and variation of the Rhopalocera of the Outer Hebrides. *Entomologist's mon. Mag.* **86**: 65–70.
300 —— 1950d. A dozen year's biogeography researches in the Inner and Outer Hebrides. *Proc. Univ. Durham phil. Soc.* **10**: 516–524.
301 —— 1953. Lepidoptera in the Isles of Lewis and Harris in 1952. *Entomologist* **86**: 53–55.
302 —— 1954a. Lepidoptera from the Outer Hebrides in 1953. *Entomologist* **87**: 83–86.
303 —— 1954b. The return of *Pararge megera* L. to County Durham. *Entomologist* **87**: 264.
304 —— 1955a. Lepidoptera noted in the Outer Hebrides in 1954. *Entomologist* **88**: 51–53.
305 —— 1955b. Lepidoptera of the lesser Skye Isles. *Entomologist's Rec. J. Var.* **67**: 141–147, 169–177.
306 —— 1956a. Immigrant Lepidoptera in the Outer Hebrides in 1955. *Entomologist* **89**: 87.
307 —— 1956b. Further observations on the Lepidoptera of the Outer Hebrides. *Entomologist* **89**: 285–290.
308 —— 1956c. Lepidoptera in the Isles of Lewis and Harris in 1955. *Entomologist's Rec. J. Var.* **68**: 40–43.
309 —— 1956d. On field studies of the distribution of the plants and animals of the Scottish Western Isles. *Proc. Linn. Soc. Lond.* **167**: 103.

310 —— 1957. Lepidoptera in the Isle of Longay. *Entomologist's Rec. J. Var.* **69**: 49.
311 —— 1958a. *Argynnis aglaia* L. in Co. Durham. *Entomologist* **91**: 32.
312 —— 1958b. Lepidoptera in the Scottish Western Isles. *Entomologist* **91**: 79–86.
313 **Heslop Harrison, J. W. & Morton, J. K.** 1952. Lepidoptera in the Isles of Raasay, Rhum, Lewis and Harris in 1951. *Entomologist* **85**: 6–13.
314 **Hesselbarth, G., Van Oorschot, H. & Wagener, S.** 1995. *Die Tagfalter der Türkei* 1–3. Bocholt, Germany.
315 **Higgs, G. E.** 1986. Three unusual butterflies in Alderney. *Bull. amat. Ent. Soc.* **45**: 155.
316 **Hirons, M. J. D.** 1994. The flora of the Farne islands. *Trans. nat. Hist. Soc. Northumb.* **57**: 69–114.
317 **Hockin, D. C.** 1981. The environmental determinants of the insular butterfly faunas of the British Isles. *Biol. J. Linn. Soc.* **16**: 63–70.
318 **Hodgson, S. B.** 1935. *Euphydryas aurinia* in west Cornwall. *Entomologist* **68**: 162–163.
319 —— 1949a. *Coenonympha tullia* at flowers [Orkney]. *Entomologist* **82**: 40.
320 —— 1949b. Migrants in Orkney. *Entomologist* **82**: 238.
321 —— 1951a. Unusual behaviour of *Polyommatus icarus*. *Entomologist* **84**: 22.
322 —— 1951b. *Pieris napi* impaled on rushes. *Entomologist* **84**: 177.
323 —— 1952a. Late flying of *Coenonympha tullia* [Orkney]. *Entomologist* **85**: 4.
324 —— 1952b. *Pieris napi* in Orkney. *Entomologist* **85**: 95.
325 [**Hogan, A. R. & Haliday, A. H.**] 1855. Notes on various insects captured or observed in the neighbourhood of Dingle, Co. Kerry in July 1854 [Great Blasket and Beginnish]. *Nat. Hist. Rev.* **2**: 50–55.
326 **Holloway, J. D.** 1980. A mass-movement of *Quercusia quercus* (L.) (Lep: Lycaenidae) in 1976. *Entomologist's Gaz.* **31**: 150.
327 **Horton, P.** 1977. Local migrations of Lepidoptera from Salisbury plain in 1976. *Entomologist's Gaz.* **28**: 281–283.
328 **Howard, G.** 1975. Lepidoptera on Hoy, Orkney. *Entomologist's Rec. J. Var.* **87**: 107–109.
329 **Howarth, T. G.** 1971a. Descriptions of a new subspecies of *Pararge aegeria* (L.) (Lep., Satyridae) and an aberration of *Cupido minimus* (F.) (Lep., Lycaenidae). *Entomologist's Gaz.* **22**: 112–118.
330 —— 1971b. The status of Irish *Hipparchia semele* L. (Lep., Satyridae) with descriptions of a new subspecies and aberrations. *Entomologist's Gaz.* **22**: 123–129.
331 **Hubbard, C. E.** 1968. *Grasses*. Harmondsworth, Middlesex: Penguin Books.
332 **Huggins, H. C.** 1928. Collecting in the Aran islands. *Entomologist* **61**: 43–44.
333 —— 1956. The Burren subspecies of *Erynnis tages* L. (Lep., Hesperiidae). *Entomologist* **89**: 241–242.
334 —— 1958. Immigrants in Tresco, Isles of Scilly, 1958. *Entomologist* **91**: 251–252.
335 —— 1959. Lepidoptera in Tresco in June. *Entomologist* **92**: 14–17, 246–249.
336 —— 1972. *Euphydryas aurinia* Rott. in the Isle of Wight. *Entomologist's Rec. J. Var.* **84**: 195–196.
337 **Hughes, H. W.** 1935. An entomologist's holiday in Aberdeen and the Shetland Isle of Unst. *Rep. Ches. ent. Soc.* 1934–36: 35–44.
338 **Hull, J. E. & Heslop Harrison, J. W.** 1937–38. The natural history of the island of Raasay. *Scott. Nat.* **1937**: 67–71, 107–113, 135–144, 145–150, 169–172; **1938**: 60–64.
339 **Institute of Biology** 1992. *Safety in biological fieldwork guidance notes for codes of practice.* London: Institute of Biology. 3rd edition.
340 [**Jeffcoate, G.**] 1994. An odd couple [mating between dingy skipper and burnet companion]. *Butterfly Conserv. News* **58**: 15.
341 **Johnson, E. E.** 1955. Difference between adjacent colonies of *Euphydryas aurinia* Rott. *Entomologist's Rec. J. Var.* **67**: 132.
342 **Jones, C. G.** 1968. Early June in Lundy: butterflies and moths recorded. *Rep. Lundy Fld Soc.* **19**: 30–33.
343 **Kane, W. F. de V.** 1907. Lepidoptera of Lambay. *Ir. Nat.* **16**: 46.
344 —— 1912. Clare island survey. *Proc. R. Ir. Acad.* **26**: 1–10.

345 **Kaye, W .J.** 1922. Lepidoptera of the smaller Channel islands. *Entomologist's Rec. J. Var.* **34**: 175–176.

346 **Kennar, R. B.** 1912. Small tortoiseshell butterfly hibernating in Shetland. *Scott. Nat.* **1912**: 92–93.

347 **Kershaw, K. A.** 1964. *Quantitative and dynamic ecology*. London: Edward Arnold.

348 **Kett, S.** 1993. *Satyrium w-album* (Knoch) (Lep., Lycaenidae) in a moth trap. *Entomologist's Rec. J. Var.* **105**: 282–283.

349 **Kevan, D. K. McE.** 1941. The insect fauna of the Isle of Eigg. *Entomologist* **74**: 247–254.

350 **King, E. B.** 1922. Lepidoptera of the smaller Channel islands. *Entomologist's Rec. J. Var.* **34**: 217.

351 —— 1923. Some notes on butterflies in England and the Channel Isles in 1922. *Entomologist* **56**: 18–19.

352 —— 1938. *D. plexippus* in the Isles of Scilly. *Entomologist* **71**: 236.

353 **King, J. J. F. X.** 1901. *Vanessa antiopa* in Shetland. *Entomologist's mon. Mag.* **37**: 226–227.

354 **Kinnear, P. K.** 1976. Unusual numbers of peacocks (*Inachis io* (L.)) (Lep., Nymphalidae) in Shetland. *Entomologist's Gaz.* **27**: 137.

355 **Kirkaldy, G. W.** 1899. *Erebia aethiops* (blandina) in the Isle of Skye. *Entomologist* **32**: 236.

356 **Kitihara, M. & Fujii, K.** 1994. Biodiversity and community structure of temperate species within a gradient of human disturbance. An analysis based on the concept of generalist vs. specialist strategies. *Res. popul. Ecol.* **36**: 187–199.

357 **Knill-Jones, S. A.** 1992. Late emergence times in the Isle of Wight during 1991. *Entomologist's Rec. J. Var.* **104**: 184.

358 **Knill-Jones, S. A. & Angell, B. J.** 1996. Exotic and rare migrant butterflies recorded on the Isle of Wight during 1995. *Entomologist's Rec. J. Var.* **108**: 143–144.

359 **Kudrna, O.** 1986. *Butterflies of Europe:* **8**. *Aspects of the conservation of the butterflies of Europe*. Wiesbaden: Aula-Verlag.

360 **Lack, D.** 1932. Further notes on insects from St Kilda in 1931. *Entomologist's mon. Mag.* **68**: 139–145.

361 **Laidlaw, W. B. R.** 1932. Speckled wood butterfly (*Pararge aegeria*) in Mull. *Scott. Nat.* **1932**: 19.

362 **Lansbury, I.** 1961. *Vanessa atalanta* in the Aran islands. *Entomologist's Gaz.* **12**: 35.

363 **Lattin, G. de** 1952. Two new subspecies of *Hipparchia semele* Linnaeus. *Entomologist's Rec. J. Var.* **64**: 335.

364 **Lavery, T. A.** 1989. The heath fritillary (*Mellicta athalia* Rott.): did it really occur in Ireland? *Bull. amat. Ent. Soc.* **48**: 158–159.

365 **Leech, M. J.** 1951. Lepidoptera in the island of South Uist, Outer Hebrides, August 17–September 4, 1950. *Entomologist* **84**: 193–194.

366 **Lekay, H. G.** 1911. *Argynnis lathonia* in Guernsey. *Entomologist* **44**: 300.

367 **Le Quesne, W .J.** 1945. Larvae of *Lampides boeticus* (Lep., Lycaenidae) in Jersey. *Entomologist's mon. Mag.* **81**: 280.

368 —— 1946. The butterflies of Jersey. *Entomologist's mon. Mag.* **82**: 22–23.

369 —— 1947. Two additions to the list of Jersey butterflies with notes on other species. *Entomologist's mon. Mag.* **83**: 134.

370 —— 1973. The development of the present state of knowledge of insects and related classes in Jersey. *A. Bull. Soc. jersiaise* **21**: 57–63.

371 **Leverton, R.** 1994. The large white, *Pieris brassicae* L. (Lep., Pieridae) apparently univoltine in Banffshire. *Entomologist's Rec. J. Var.* **106**: 190–191.

372 **Long, R. C.** 1970. Rhopalocera (Lep.) of the Channel islands. *Entomologist's Gaz.* **21**: 241–251.

373 —— 1987. Two butterfly records from Jersey, Channel Islands. *Entomologist's Gaz.* **38**: 202.

374 **Longstaff, G. B.** 1907. First notes on the Lepidoptera of Lundy Island. *Entomologist's mon. Mag.* **43**: 241–244.

375 **Lorimer, R. I.** 1981. Orkney Lepidoptera. *Entomologist's Gaz.* **32**: 75–78.

376 —— 1983. *The Lepidoptera of the Orkney Islands.* Faringdon, Oxon: E. W. Classey.

377 —— 1988. The Lepidoptera of the Orkney Islands: supplement, 1983–87. *Entomologist's Gaz.* **39**: 181–186.

378 **Lowe, F. E.** 1904. *Lampides boeticus* in the Channel Islands. *Entomologist's Rec. J. Var.* **16**: 297.
379 —— 1911. *Lampides boeticus* in Guernsey. *Entomologist* **44**: 367.
380 **Luff, W. A.** 1873. A list of the butterflies inhabiting Guernsey and Sark, with notes on their occurrence. *Entomologist* **6**: 324–326.
381 —— 1874. Additions to the list of Macro-Lepidoptera inhabiting Guernsey and Sark. *Entomologist* **7**: 42–43; **8**: 29–32.
382 —— 1882. Butterflies of Guernsey and Sark. *Rep. Trans. Guernsey Soc. nat. Sci.* **1**: 61.
383 —— 1886. *Anosia plexippus* in Guernsey. *Entomologist* **19**: 278.
384 —— 1890a. *Hesperia lineola* in Jersey. *Entomologist* **23**: 93.
385 —— 1890b. Captures on the Isle of Jethou. *Entomologist's Rec. J. Var.* **1**: 139–140.
386 —— 1893a. *Catocala fraxini, Colias edusa* and *hyale* in Guernsey. *Entomologist's Rec. J. Var.* **4**: 298.
387 —— 1893b. *Anthocharis cardamines* in Guernsey. *Entomologist's mon. Mag.* **29**: 139.
388 —— 1898. List of insects on Alderney. *Rep. Trans. Guernsey Soc. nat. Sci.* **3** (1897): 175–182.
389 —— 1900a. The insects of Alderney. *Rep. Trans. Guernsey Soc. nat.* Sci. **3** (1899): 388–408.
390 —— 1900b. *Papilio machaon* and *Lampides boeticus* in Guernsey. *Entomologist's Rec. J. Var.* **12**: 273.
391 —— 1905a. The insects of Jethou. *Rep. Trans. Guernsey Soc. nat. Sci.* **4** (1904): 388–390.
392 —— 1905b. The insects of Herm. *Rep. Trans. Guernsey Soc. nat. Sci.* **4** (1904): 374–378.
393 —— 1907. The insects of Sark. Rep. *Trans. Guernsey Soc. nat. Sci.* **5** (1906): 185–199.
394 **Lupton, H.** 1880. The Lepidoptera of Arran. *Naturalist, Hull* **6**: 72–75.
395 **MacArthur, R. H. & Wilson, E. O.** 1967. The theory of island biogeography. *New Jersey: Princeton University Press.*
396 **MacGillivray, J.** 1842. An account of the island of St Kilda, chiefly with reference to its natural history, made during a visit in 1840. *Edinb. new phil. J.* **32**: 47–70.
397 **Majerus, M. E. N.** 1979. The status of *Anthocharis cardamines hibernica* Williams (Lepidoptera: Pieridae), with special reference to the Isle of Man. *Entomologist's Gaz.* **30**: 245–248.
398 —— 1980. The brimstone butterfly in the Isle of Man. *Bull. amat. Ent. Soc.* **39**: 40–41.
399 **Martineau, A. H.** 1894. [Exhibition – of Lepidoptera collected on Lundy by R. W. Chase at Birm. Ent. Soc.]. *Entomologist* **27**: 303.
400 **Marwick, J. G.** 1931. Small tortoiseshell butterfly in Orkney. *Scott. Nat.* **1931**: 89.
401 **Matthews, L. H.** 1932. Denny Isle. *Proc. Bristol Nat. Soc.* **7**: 371–378.
402 **McAllister, D. W.** 1993. The spread of the speckled wood *Pararge aegeria* (L.) in Easter Ross and Sutherland. *Butterfly Conserv. News* **53**: 44–48.
403 **[McArthur, H.]**. 1901. Four months collecting in the Isle of Lewis. *Entomologist* **34**: 305–306.
404 **McLeod, L.** 1972. The distribution of insects related to railway embankments. *Entomologist's Rec. J. Var.* **84**: 69–71.
405 **Mere, R. M.** 1959. Isles of Scilly in 1958. *Entomologist's Gaz.* **10**: 107–110.
406 —— 1960. Isles of Scilly in 1959. *Entomologist's Gaz.* **11**: 118.
407 **Miller, K. W. & Owen, J. A.** 1952. A list of insects from the Isle of Ulva. *Scott. Nat.* **1952**: 31–37.
408 **Morgan, M. J.** 1969. A check list of insects from Bardsey Island, Caernarvonshire. *Entomologist's Gaz.* **20**: 105–117.
409 **Morton, A. C. G.** 1982. The effects of marking and capture on recapture frequencies of butterflies. *Oecologia* **53**: 105–110.
410 **Muir, D. A.** 1954. A note on the insects from the Garvellach Islands. *Glasg. Nat.* **17**: 140–141.
411 **Murdoch, D., Lyle, T. & Holmes, M. J. G.** 1987. The butterflies of London's East End: a three year survey. *Lond. Nat.* **66**: 125–133.
412 **Nässig, W. A.** 1995. Die Tagfalter der Bundesrepublic Deutschland: Vorschlag für ein modernes, phylogenetisch orientiertes Artenverzeichnis (kommentierte Checkliste) (Lepidoptera, Rhopalocera). *Ent. Nachr. Ber.* **39**: 1–28.
413 **Nelson, J. M.** 1980. Some invertebrates from Ailsa Craig. *W. Nat.* **1980**: 9–16.
414 **Newman, L. H.** 1931. Sark Lepidoptera. *Entomologist's Rec. J. Var.* **43**: 184–186.
415 **Nicholson, C.** 1933. Immigrant Lepidoptera [Round Island]. *Entomologist* **66**: 285–286.
416 **Nonweiler, T.** 1949. *Danaus plexippus* on Skye. *Entomologist* **82**: 40.

417 **Norgate, F.** 1880. List of insects observed in Tresco, Scilly Isles, in August 1878. *Entomologist's mon. Mag.* **16**: 182–183.

418 **Nye, E. R.** 1957. Insect records from Bardsey. *Entomologist's Rec. J. Var.* **69**: 206–208.

419 **Oates, M. R. & Warren, M. S.** 1990. *A review of butterfly introductions in Britain and Ireland.* London: WWF.

420 **Oldham, C.** 1931. Painted lady in Shetland. *Scott. Nat.* **1931**: 152.

421 **Owen, D. F.** 1949. The Macrolepidoptera of the Moorgate, London, bombed sites. *Entomologist* **82**: 59–62.

422 —— 1951. Bombed site Lepidoptera. *Entomologist* **84**: 265–272.

423 **Pankhurst, R. J. & Allinson, J.** 1985. *British grasses: a punched-card key to grasses in the vegetative state.* AIDGAP Guides. Occasional Publication 10.

424 **Parrack, J. D.** 1986. Entomological investigation of the 'Snook', Holy Island, part of the Lindisfarne NNR, during 1984–86. *Vasculum* **71**: 20–29.

425 **Parsons, T.** 1978. An insect island. In Allsop, K. & Fowles, J. (Eds) *Steepholm – A case history in the study of evolution.* Kenneth Allsop Mem. Trust, Dorset. pp. 177–187.

426 **Peacock, A. D., Smith, E. P. & Davidson, C. F.** (Eds) 1934–35. The natural history of South Rona. The results of a biological expedition from University College (St Andrew's, Dundee), July 1933. *Scott. Nat.* **1934**: 113–127, 149–163; **1935**: 3–10, 31–34.

427 **Pennington, M. G.** 1993. Butterflies in Shetland. *BBCS News* **55**: 45–47.

428 —— 1996a. Camberwell Beauties *Nymphalis antiopa* (L.) (Lep. Nymphalidaee) and Swallowtails *Papilio machaon* L. (Lep. Papilionidae) in Shetland. *Entomologist's Rec. J Var.* **108**: 67–68.

429 —— 1996b. Large White butterfly *Pieris brassicae* (L.) (Lep.: Pieridae) in Shetland. *Entomologist's Rec. J. Var.* **108**: 137–139.

430 —— 1996c. The Peacock butterfly *Inachis io* (L.) (Lep. Nymphalidae) in Shetland. *Entomologist's Rec. J. Var.* **108**: 208–210.

431 **Pennington, M. & Riddiford, N.** 1993. Moths and butterflies in Shetland in 1992. *Shetland Entomological Group. Supplement to Newsletter* no. 4: 2–8.

432 **Piquet, F. G.** 1873. A list of the butterflies inhabiting Jersey, with notes of their occurrence. *Entomologist* **6**: 399–401.

433 **Plant, C.** 1987. *The butterflies of the London area.* London: London Natural History Society, Passmore Edwards Museum.

434 **Pollard, E.** 1979. Population ecology and change in range of the white admiral butterfly *Ladoga camilla* L. in England. *Ecol. Ent.* **4**: 61–74.

435 **Pollard, E., Hall, M. L. & Bibby, T .J.** 1986. Monitoring the abundance of butterflies 1976-85. *Research and Survey in Nature Conservation Series* no. 2. Peterborough: Nature Conservancy Council.

436 **Pollard, E., Moss, D. & Yates, T .J.** 1995. Population trends of common British butterflies at monitored sites. *J. appl. Ecol.* **32**: 9–16.

437 **Pollard, E. & Yates, T .J.** 1992. The extinction and foundation of local butterfly populations in relation to population variability and other factors. *Ecol. Ent.* **17**: 249–254.

438 —— 1993. *Monitoring butterflies for ecology and conservation.* London: Chapman & Hall.

439 **Pratt, C.** 1986–87. A history and investigation into the fluctuations of *Polygonia c-album* L. the comma butterfly. *Entomologist's Rec. J. Var.* **98**: 197–203, 244–250; **99**: 21–27, 69–80.

440 **Quinette, J. & Lepertel, N.** 1993. Inventaire préliminaire des Macrolépidoptères du département de la Manche. *Alexanor* **18**: 157–177.

441 **Ralston, G. S.** 1959. Migrating butterflies in the northern isles of Shetland. *Entomologist* **92**: 184.

442 —— 1960. Migrating butterflies in the northern isles of Shetland. *Entomologist* **93**: 194.

443 **Ravenscroft, N. O. M.** 1994a. The ecology of the chequered skipper butterfly *Carterocephalus palaemon* in Scotland. I. Microhabitat. *J. appl. Ecol.* **31**: 613–622.

444 —— 1994b. The ecology of the chequered skipper butterfly *Carterocephalus palaemon* in Scotland. II. Foodplant quality and population range. *J. appl. Ecol.* **31**: 623–630.

445 **Redway, D. B.** 1981. Some comments on the reported occurrence of *Erebia epiphron* (Knoch) (Lepidoptera: Satyridae) in Ireland during the nineteenth century. *Entomologist's Gaz.* **32**: 157–159.

446 **Reed, T. M.** 1985. The number of butterfly species on British islands. *Proc. 3rd Congr. Eur. Lepid., Cambs.* **1982**: 146–152.
447 **Rees, C. J. & Sutton, S. L.** 1960. Entomological survey, Cape Clear Island 1960. Interim report. *Cape Clear Bird Obs. Rep.* **2**: 29–30.
448 **Richardson, A.** 1960. A week in the Isle of Arran. *Entomologist's Rec. J. Var.* **72**: 112–115.
449 —— 1963. Further observations on the Lepidoptera of the Isles of Scilly with particular reference to St Agnes. *Entomologist's Rec. J. Var.* **75**: 179–184.
450 **Richardson, A. & Mere, R. M.** 1958. Some preliminary observations on the Lepidoptera of the Isles of Scilly, with particular reference to Tresco. *Entomologist's Gaz.* **9**: 115–147.
451 **Riley, N. D.** 1975. Jersey Lepidoptera. *Entomologist* **58**: 149–151.
452 **Rintoul, L. J. & Baxter, E. V.** 1920. Lepidoptera of the Isle of May. Scott. Nat. 1920: 198.
453 —— 1937. Peacock butterflies on Islay. *Scott. Nat.* **1937**: 174.
454 —— 1950. Natural history notes from the Isle of Gigha. *Scott. Nat.* **1950**: 93–97.
455 **Rockingham, N. W.** 1949. *Pieris napi* in Orkney. *Entomologist* **82**: 272.
456 **Rodwell, J. S.** (Ed.) 1991–95. *British plant communities* **1–5**. Cambridge: Cambridge University Press.
457 **Rogers, H. M.** 1944. Calf Island. Some natural history notes for 1943. *Peregrine* **1**: 12–14.
458 **Roland, J., McKinnon, G., Backhouse, C. & Taylor, P .D.** 1996. Even smaller radar tags on insects. *Nature, Lond.* **381**: 120.
459 **Rushton, D. R. A.** 1971. A peacock butterfly in Shetland. *Entomologist's Rec. J. Var.* **83**: 397.
460 **Scoble, M. J.** 1992. *The Lepidoptera. Form, function and diversity.* Oxford: Oxford University Press.
461 **Scott, J. A.** 1985. The phylogeny of butterflies (Papilionoidea and Hesperioidea). *J. Res. Lepid.* **23**: 241–281.
462 **Shapiro, A. M.** 1970. Notes on the spring butterflies in north-east Scotland and Orkney. *Entomologist's Rec. J. Var.* **82**: 85–86.
463 **Shaw, M. W.** 1951. *Danaus plexippus* in the Hebrides. *Entomologist* **84**: 279.
464 **Shayer, C. J.** 1947. *Argynnis selene* in Sark. *Entomologist* **80**: 179.
465 —— 1967. List of butterflies of the Guernsey Bailiwick. *Rep. Soc. Guernes* **18**: 28–29.
466 **Shreeve, T. G.** 1992. Monitoring butterfly movements. pp. 120–138. In Dennis, R. L. H. (Ed.) *The ecology of butterflies in Britain.* Oxford: Oxford University Press.
467 —— 1995. Butterfly mobility. pp. 37–45. *In* Pullin, A. S. (Ed.), *Ecology and conservation of butterflies.* London: Chapman & Hall.
468 **Shreeve, T. G., Dennis, R. L. H. & Pullin, A. S.** 1996. Marginality: scale determined processes and the conservation of the British butterfly fauna. *Biodiversity & Conservation,* **5**: 1131–1141.
469 **Shreeve, T. G., Dennis, R. L. H. & Williams, W. R.** 1995. Uniformity of wing spotting of *Maniola jurtina* (L.) (Lepidoptera: Satyrinae) in relation to environmental heterogeneity. *Nota lepid.* **18**: 77–92
470 **Simmons, I. G. & Tooley, M. J.** 1981. *The environment in British prehistory.* London: Duckworth.
471 **Simpson, M. S. L.** 1974. Butterflies on the Isles of Scilly, June 1973. *Bull. amat. Ent. Soc.* **33**: 91–96.
472 **Skinner, B.** 1984. *Colour identification guide to moths of the British Isles.* Middlesex: Viking.
473 **Smith, A. A. D.** 1933. *Danaus plexippus* in the Scilly Isles. *Entomologist* **66**: 6.
474 **Smith, K. G. V.** 1958. Some records of Diptera and other insects from Lundy. *Entomologist's mon. Mag.* **94**: 94.
475 —— 1960. Notes on some insects from Caldey island. *Nature Wales* **6**: 46–48.
476 **Smith, K. G. V. & Smith, V.** 1983. *A Bibliography of the Entomology of the smaller British offshore Islands.* Faringdon, Oxon: Classey.
477 **Smith, M. J. M.** 1984. Camberwell beauty in Shetland – 1982. *Bull. amat. Ent. Soc.* **43**: 146–147.
478 **[Smith, R.]** 1993. Note on holly blue *Celastrina argiolus* (L.) (Lycaenidae) at Spurn Head in 1992. *Butterfly Conserv. News* **53**: 22.
479 **Smout, A. M. & Kinnear, P.** 1993. *The butterflies of Fife.* Glenrothes, Fife: Fife Nature.

480 **Sneath, P. H. A. & Sokal, R. R.** 1973. *Numerical taxonomy*. San Francisco, W. H. Freeman & Co.
481 **South, R.** 1888a. Lepidoptera of the Outer Hebrides. *Entomologist* **21**: 25–27.
482 —— 1888b. Distribution of Lepidoptera in the Outer Hebrides, Orkney and Shetland. *Entomologist* **21**: 28–30, 98–99.
483 —— 1893. Lepidoptera of the Shetland Islands. *Entomologist* **26**: 98–102.
484 **STATISTICA.** 1994. *STATISTICA for windows* **1–3**. Tulsa, OK: Statsoft, Inc.
485 **Stewart, A. M.** 1925. Notes from Arran. *Entomologist* **58**: 272–273.
486 —— 1933. *Colias croceus* in Arran. *Entomologist* **66**: 228.
487 —— 1939. *Nymphalis io* in Arran. *Entomologist* **72**: 8.
488 **Stewart, M.** 1950. *Pieris napi* in Orkney. *Entomologist* **83**: 43–44.
489 **Sullivan, D. J.** 1946. Remarkable migration of butterflies at night and during a gale in Co. Donegal. *Ir. Nat. J.* **8**: 337.
490 **Summers, G.** 1975a. Butterflies in the Isles of Scilly. *Entomologist's Rec. J. Var.* **87**: 95–96.
491 —— 1975b. Butterflies on Looe (St George's) Island. *Entomologist's Rec. J. Var.* **87**: 278.
492 —— 1976a. Lepidoptera in the Isles of Scilly and Cornwall, June 1975. *Entomologist's Rec. J. Var.* **88**: 29–30.
493 —— 1976b. Late September in Scilly. *Entomologist's Rec. J. Var.* **88**: 239.
494 —— 1977a. October in Scilly. *Entomologist's Rec. J. Var.* **89**: 18.
495 —— 1977b. Treshnish Isles. *Entomologist's Rec. J. Var.* **89**: 173.
496 **Tait, W.** 1878. *Colias edusa* in Orkney. *Scott. Nat.* **1878**: 160.
497 **Thomas, C. D.** 1985. Specializations and polyphagy of *Plebejus argus* (Lepidoptera: Lycaenidae) in North Wales. *Ecol. Ent.* **10**: 325–340.
498 —— 1994. Extinction, colonization and metapopulations: environmental tracking by rare species. *Conserv. Biol.* **8**: 373–378.
499 —— 1995. Ecology and conservation of butterfly metapopulations in the fragmented British landscape. pp. 46–63 *In* Pullin, A. S. (Ed.) *Ecology and conservation of butterflies*. London: Chapman & Hall.
500 **Thomas, C. D. & Jones, T. M.** 1993. Partial recovery of a skipper butterfly (*Hesperia comma*) from population refuges: lessons for conservation in a fragmented landscape. *J. anim. Ecol.* **62**: 472–481.
501 **Thomas, C. D., Thomas, J. A. & Warren, M. S.** 1992. Distributions of occupied and vacant habitats in fragmented landscapes. *Oecologia* **92**: 563–567.
502 **Thomas, J. A.** 1983a. The ecology and conservation of *Lysandra bellargus* (Lepidoptera: Lycaenidae) in Britain. *J. appl. Eeol.* **20**: 59–83.
503 —— 1983b. A quick method for estimating butterfly numbers during surveys. *Biol. Conserv.* **27**: 195–211.
504 —— 1984. The conservation of butterflies in temperate countries; past efforts and lessons for the future. pp. 333–353. *In* Vane-Wright, R. I. and Ackery, P. R. (Eds) *The Biology of Butterflies*. London: Academic Press.
505 —— 1986. *RSNC Guide to Butterflies of the British Isles*. Middlesex: Country Life Books, Newnes
506 **Thomas, J. A., Moss, D. & Pollard, E.** 1994. Increased fluctuation of butterfly populations towards the northern edges of species' ranges. *Ecography* **17**: 215–220.
507 **Thomas, J. A. & Webb, N.** 1984. *Butterflies of Dorset* (includes Brownsea Island). Dorchester: The Dorset Natural History & Archaeological Society.
508 **Thomson, G.** 1970. On the nature of *Maniola jurtina splendida* B. White (Lep., Satyridae). *Entomologist's Rec. J. Var.* **82**: 261–268.
509 —— 1971. The Manx race of *Maniola jurtina* (L.) (Lep., Satyridae). *Entomologist's Rec. J. Var.* **83**: 91–94.
510 —— 1973. Temperature effects on *Maniola jurtina* (L.) (Lep., Satyridae). *Entomologist's Rec. J. Var.* **85**: 109–114.
511 —— 1980. *Butterflies of Scotland*. London: Croom Helm.
512 —— 1987. *Enzyme variation at morphological boundaries in* Maniola *and related genera*. Ph.D. Thesis, University of Stirling.

513 **Traill, J. W. H.** 1869. Notes on the Lepidoptera of Orkney. *Entomologist* **4**: 197–200.
514 —— 1888. The Lepidoptera of the Outer Hebrides, Orkney and Shetland. *Scott. Nat.* **1888**: 298–304.
515 **Tremewan, W. G.** 1953. *Aglais urticae* flying by night. *Entomologist* **86**: 58.
516 —— 1995. The genus *Zygaena* Fabricius (Lepidoptera, Zygaenidae) on St Michael's Mount, Cornwall. *Entomologist's Gaz.* **46**: 256.
517 **Trevor, J.** 1994. Butterflies of the Outer Hebrides. *Butterfly Conservation News* **58**: 30–31.
518 **Turner, H. J.** 1924. Divergence of character [*Aglais urticae* on Man]. *Entomologist's Rec. J. Var.* **36**: 156.
519 **Turner, J. R. G.** 1986. Why are there so few butterflies in Liverpool? Homage to Alfred Russel Wallace. *Antenna* **10**: 18–24.
520 **Turner, J. R. G., Gatehouse, C. M. & Corey, C. A.** 1987. Does solar energy control organic diversity? Butterflies, moths and the British climate. *Oikos* **48**: 195–205.
521 **Tutt, J. W.** 1894. Societies [Lundy records]. *Entomologist's Rec. J. Var.* **5**: 105–111.
522 **Vickery, M.** 1991. National garden butterfly survey. *Butterfly Conserv. News* **47**: 32–38.
523 —— 1993. Garden butterfly survey 1992. *Butterfly Conserv. News* **53**: 50–59.
524 **Vine-Hall, J. W.** 1969. Some notes on Lepidoptera from the Hebrides. *Entomologist's Gaz.* **20**: 53–58.
525 **Wagener, P .S.** 1988. What are the valid names for the two genetically different taxa included within *Pontia daplidice* (Linnaeus, 1758)? (Lepidoptera: Pieridae). *Nota lepid.* **11**: 21–38.
526 **Walker, D.** 1970. Direction and rate in some British post-glacial hydroseres. pp. 117–139. *In* Walker, D. & West, R. G. (Eds) *Studies in the vegetational history of the British Isles*. Cambridge: Cambridge University Press.
527 **Walker, F. A.** 1878. *Catalogue of Lepidoptera Rhopalocera in the collection of the Rev. F. A. Walker, M.A., F.L.S.* London: T P. Newman.
528 —— 1888. Lepidoptera in the Channel islands. *Entomologist* **21**: 150–151.
529 —— 1892. Are Jersey insects British? *Entomologist* **25**: 320.
530 —— 1897. Entomology in Alderney. *Entomologist* **30**: 223.
531 **Walker, J. J.** 1922. On the occasional occurrence of butterflies in Iceland; with notes on the lepidopterous fauna of the North Atlantic islands. *Entomologist's mon. Mag.* **58**: 1–7.
532 —— 1932. Some records of insects from St Kilda. *Entomologist's mon. Mag.* **68**: 146–150.
533 **Walmsley, T. & Warlsley, P.** 1977. [The feeding habits of butterflies]. *Rep. Lundy Fld Soc.* **28**: 44.
534 **Warren, M. S.** 1987b. The ecology and conservation of the heath fritillary, *Mellicta athalia*. I. Host selection and phenology. *J. appl. Ecol.* **24**: 467–482.
535 —— 1987b. The ecology and conservation of the heath fritillary, *Mellicta athalia*. II. Adult population structure and mobility. *J. appl. Ecol.* **24**: 483–498.
536 —— 1987c. The ecology and conservation of the heath fritillary, *Mellicta athalia*. III. Population dynamics and the effect of habitat management. *J. appl. Ecol.* **24**: 499–513.
537 —— 1992. Butterfly populations. pp. 73–92. *In* Dennis, R. L. H. (Ed.) *The Ecology of Butterflies in Britain*. Oxford: Oxford University Press.
538 **Warren, M. S., Pollard, E. & Bibby, T. J.** 1986. Annual and long-term changes in a population of the wood white butterfly *Leptidea sinapis*. *J. anim. Ecol.* **55**: 707–719.
539 **Waterston, J.** 1906. On some invertebrates from St Kilda. *Ann. Scot. Nat. Hist.* **1906**: 150–153.
540 —— 1911. *Vanessa cardui* in North Mavine, Shetland. *Entomologist's mon. Mag.* **47**: 217.
541 —— 1913. *Pyrameis atalanta* in North Mavine (Shetland). *Entomologist's mon. Mag.* **49**: 214.
542 —— 1939. Insects from Colonsay, South Ebudes. *Scott. Nat.* **1939**: 128–131.
543 **Watson, A. B.** 1893. Collecting in Arran. *Entomologist* **26**: 52–54.
544 **Weir, J. J.** 1880. The Macrolepidoptera of the Shetland Isles. *Entomologist* **13**: 249–251, 289–291.
545 —— 1881a. Notes on the Lepidoptera of the Outer Hebrides. *Entomologist* **14**: 218–223.
546 —— 1881b. The Macrolepidoptera of the Shetland Isles. *Entomologist* **14**: 278–281.
547 —— 1882a. Notes on the Lepidoptera of the Orkney Islands. *Entomologist* **15**: 1–5.
548 —— 1882b. The Macrolepidoptera of the Isle of Arran. *Entomologist* **15**: 250–253.
549 —— 1884. The Macrolepidoptera of the Shetland Isles. *Entomologist* **17**: 1–4.

550 **Wheeler, G.** 1931a. *Melitaea aurinia* Rott.: its fluctuation and variation. *Entomologist* **64**: 137.
551 —— 1931b. The flying powers of *Melitaea aurinia*. *Entomologist* **64**: 223.
552 **White, B.** 1882. The Lepidoptera of Orkney, Shetland and the Outer Hebrides. *Scott. Nat.* **1882**: 289–291, 337–344.
553 **White, F. B. W.** 1872. Insecta Scotica. *Scott. Nat.* **1872**: 162–168, 198–202, 238–241, 273–275.
554 **White, P. J.** 1892–93. Entomology. Report for the Years 1892–93. *Puffin Island Biological Station*: 10–11.
555 **Wild, O. H.** 1922. Breeding of red admiral in Bute. *Scott. Nat.* **1922**: 27.
556 **Wilks, R. L.** 1941. Lepidoptera in the Isle of Islay. *Entomologist* **74**: 185–186.
557 —— 1945a. *Argynnis cydippe* in Islay [probably error for *Argynnis aglaia*]. *Entomologist* **78**: 86.
558 —— 1945b. Some notes on the Macrolepidoptera of Islay. *Entomologist* **78**: 161–165.
559 **Wilkinson, R. S.** 1975. The scarce swallow-tail: *Iphiclides podalirius* (L.) in Britain. *Entomologist's Rec. J. Var.* **87**: 289–293.
560 —— 1982. The Scarce Swallow-tail: *Iphiclides podalirius* (L.) in Britain. II: Haworth's Prodromus and Lepidoptera Britannica. *Entomologist's Rec. J. Var.* **94**: 168–172.
561 **Williamson, M.** 1981. *Island populations*. Oxford: University Press.
562 **Wollaston, T. V.** 1845. Note on the entomology of Lundy island. *Zoologist* **3**: 897–900.
563 **Woof, W. E.** 1958. *An Ecological Survey of the Insects of the Farne Islands*. Ph.D. Thesis, University of Durham.
564 **Wormell, P.** 1982. The entomology of the Isle of Rhum NNR. *Biol. J. Linn. Soc.* **18**: 291–401.
565 —— 1983. Lepidoptera in the Inner Hebrides. *Proc. R. Soc. Edinb.* **83B**: 531–546.
566 **Worms, C. G. M. de** 1939. An entomological trip to Shetland, July 1938. *Entomologist* **72**: 60–65.
567 —— 1959. Lepidoptera in the Highlands and Shetlands, August 1958. *Entomologist* **72**: 1–5.
568 —— 1960. *Colias croceus* (Lep.) on Tresco and near the Lizard. *Entomologist's mon. Mag.* **93**: 60.
569 —— 1969. *Pyrameis cardui* and other butterflies in Orkney. *Entomologist's Rec. J. Var.* **81**: 282.
570 —— 1970. Two weeks in Orkney, July-August 1969. *Entomologist's Rec. J. Var.* **82**: 67–74.
571 —— 1972. Collecting Lepidoptera in Britain during 1971. *Entomologist's Rec. J. Var.* **84**: 184–193.

APPENDIX 1

Butterflies occurring on the British and Irish islands as rare immigrants and accidental and deliberate introductions.

(Adapted from Emmet & Heath, 1989; see also Knill-Jones & Angell, 1996.)

R = rare immigrant
D = Deliberate introduction
A = Accidental introduction, adventive, escape from captivity
? = status (identification and mode of entry) uncertain

Superfamily HESPERIOIDEA

Family Hesperiidae

Subfamily Hesperiinae

Hylephila phyleus (DRURY, 1773) Fiery Skipper. (A)

Subfamily Pyrginae

Carcharodus alceae (ESPER, 1780) Mallow Skipper (A)
Pyrgus armoricanus (OBERTHÜR, 1910) Oberthür's Grizzled Skipper (A)

Superfamily PAPILIONOIDEA

Family Papilionidae

Subfamily Parnassiinae

Parnassius apollo (LINNAEUS, 1758) The Apollo (R but also A & D)
Parnassius phoebus (FABRICIUS, 1793) Small Apollo (A)

Subfamily Zerynthiinae

Zerynthia rumina (LINNAEUS, 1758) Spanish Festoon (A)
Zerynthia polyxena ([DENIS & SCHIFFERMÜLLER], 1775) (D or A)

Subfamily Papilioninae

Papilio glaucus LINNAEUS, 1758 Tiger Swallowtail (A)
Papilio demetrius CRAMER, 1782 Black Swallowtail (A)

Family Pieridae

Subfamily Coliadinae

Colias palaeno (LINNAEUS, 1761) Moorland Clouded Yellow (A)
Colias alfacariensis BERGER, 1948 Berger's Clouded Yellow (R)
Gonepteryx cleopatra (LINNAEUS, 1767) The Cleopatra (R, D & A)

Subfamily Pierinae

Euchloe crameri (BUTLER, 1869) Butler's Dappled White (R), or *E. ausonia* (HÜBNER, [1804]), not *E. simplonia* (FREYER) as in Emmet & Heath (1989)

Family Lycaenidae

Subfamily Theclinae

Rapala schistacea (MOORE, [1881]) Slate Flash (A)

Subfamily Lycaeninae

Lycaena tityrus (PODA, 1761) Sooty Copper (R)
Lycaena alciphron (ROTTEMBURG, 1775) Purple-shot Copper (A)

Subfamily Polyommatinae

Leptotes pirithous (LINNAEUS, 1767) Lang's Short-tailed Blue (R)
Plebicula dorylas ([DENIS & SCHIFFERMÜLLER], 1775) Turquoise Blue (? A)
Glaucopsyche alexis (PODA, 1761) Green-underside Blue (A or R)

Family Nymphalidae

Subfamily Heliconiinae

Dryas julia FABRICIUS, 1775 The Julia (A)
Heliconius charitonius (LINNAEUS, 1767) (A)

Subfamily Nymphalinae

Junonia villida (FABRICIUS, 1787) Albin's Hampstead Eye (? A)
Junonia oenone (LINNAEUS, 1758) Blue Pansy (A)
Colobura dirce (LINNAEUS, 1758) The Zebra (A)
Hypanartia lethe (FABRICIUS, 1793) Small Brown Shoemaker (A)
Vanessa indica (HERBST, 1794) Indian Red Admiral (R or A)
Nymphalis xanthomelas ([DENIS & SCHIFFERMÜLLER], 1775) Scarce Tortoiseshell (R)
Araschnia levana (LINNAEUS, 1758) European Map (D & A)

Subfamily Argynninae

Boloria dia (LINNAEUS, 1767) Weaver's Fritillary (A, D)
Argynnis aphrodite (FABRICIUS, 1787) Aphrodite Fritillary (A)
Argynnis niobe (LINNAEUS, 1758) Niobe Fritillary (possibly extinct resident, but also D or R)
Argynnis pandora ([DENIS & SCHIFFERMÜLLER], 1775) Mediterranean Fritillary (R)
Phalanta phalantha (DRURY, [1773]) Common Leopard (A)

Subfamily Melitaeinae

Melitaea didyma (ESPER, 1779) Spotted Fritillary (A)

Subfamily Satyrinae

Lasiommata maera (LINNAEUS, 1758) Large Wall (A)
Erebia alberganus (DE PRUNNER, 1798) Almond-eyed Ringlet (?)
Hipparchia fagi (SCOPOLI, 1763) Woodland Grayling (A)
Chazara briseis (LINNAEUS, 1764) The Hermit (? A)
Arethusana arethusa ([DENIS & SCHIFFERMÜLLER], 1775) False Grayling (A or R)

APPENDIX 2

Observation of butterflies on islands

A. A guide to making observations on butterfly and moth species during short visits to islands

These notes are intended only as a brief guide of how and what to observe and record during short visits to islands. Much of it is obvious and common sense, but nonetheless it may prove to be a useful reminder of what to do for those who intend to plan for their visit. Needless to say all records are useful, even those made casually without preparation. The issues addressed and the points made do not differ in substance from those associated with local recording, for example recording for county atlases. However, emphasis is placed on determining the breeding status of species observed on islands.

a. Before visits

It is particularly useful to note species previously found on the island to be visited. The present list used in conjunction with the bibliography and a standard systematic text (e.g., Thomas, 1986; Emmet & Heath, 1989) should indicate which species are most likely to be observed as adults on a visit.

It is also useful to make a note of additional species recorded within the 100 km square or so of the nearest faunal source on the mainland, as well as on adjacent islands (Emmet & Heath, 1989).

Familiarize yourself with the features (i.e., wing patterns) and biology of species likely to be encountered on visits. In particular, take note of butterfly or moth species that could be confused with butterflies you may encounter (see section I.1D). Also note typical habitats, hostplants and details which will help you to find and recognise early stages of each species (see Dickson, 1992).

b. During visits

Equipment

At very least take with you a detailed map (OS 1 : 25,000 scale), a note book, an identification text for butterflies (e.g., Thomas, 1986), flowering plants, grasses and sedges (e.g., Hubbard, 1968; Fitter, Fitter & Blamey, 1978; Fitter, Fitter & Farrer, 1984; Pankhurst & Allinson, 1985), a pocket lens (×10 magnification) and a folding net for checking identifications. Fine felt-tip permanent markers (various colours) are invaluable if more precise population estimates are to be made using mark-release-recapture methods. A camera is essential equipment.

Various accessories (flash unit; tele-macro lens) can greatly assist in obtaining a valid record for a species. It is particularly useful to have the camera pre-set for a close-up shot of about a metre or so, though this will depend on the lenses being used. Closer approaches can be made once a photograph has been taken for reference at a distance unlikely to disturb the butterfly. A pair of binoculars (low power; wide field of view) is also important for identifying species in inaccessible locations and observing behaviour without causing disturbance. Details on cameras and photography can be found in Angel (1975). It is not necessary to kill or collect British butterflies to identify them. Most moths can also be identified against a good reference text (e.g., Skinner, 1984), but unfortunately some can only be determined to species' level by dissection.

Conditions

Take note of the time at the start of your observations during your visit. Note down the weather conditions, in particular the duration of bright sunshine and cloud, and a record of windspeed. The Beaufort scale is useful for the latter. If weather conditions are being monitored on the island, then obtain a record of the shade temperatures during the visit.

Cover

For small islands, try to visit as much of the island as possible. For large islands, visit as many different habitats as possible. If you have obtained a map at a scale of 1 : 25,000 or larger then you can plan your route. However, it is necessary to emphasize some important points regarding visits to islands as many of them are potentially hazardous; therefore we recommend that the field work guidelines of the Institute of Biology (1992) are followed.

Observations

If you are uncertain about an observation, then take a photograph as a record, and have unusual identifications confirmed by colleagues who are with you.

A map will allow you to give pin-point references to locations on the island and enable you to divide your route into sections much as the transects carried out in the ITE Butterfly Monitoring Scheme (see Pollard & Yates, 1993). Make notes on the habitats. Distinct habitats can be sketched on the map. Then, in years to come, new records can be related to any changes in habitat structure. Simple methods exist for the subjective assessment of plant abundance and for describing habitat structure and composition, such as the Braun-Blanquet and Domin techniques (see Kershaw, 1964) and the national vegetation communities classification (Rodwell, 1991–95).

Count numbers of adults seen of both sexes separately. If you make a note of your route, your visit becomes a transect and you can calculate the distance covered and the area over which observations have been made, as well as the time spent in each habitat.

Make observations on all aspects of butterfly behaviour, but especially those activities that allow determination of the breeding status of species, such as:

- teneral adults, that is adults that have recently eclosed and are still drying their wings;
- state of wing wear of individuals seen; a simple four-state scoring scale (e.g., 1, fresh; 2, no wear but not fresh; 3, significant scale loss and wing-edge tattering; 4, worn and tattered) can help classify individuals seen;
- mating behaviour, such as territorial disputes (i.e., spiral interactions; horizontal chases), attempted courtships or copulation;
- egglaying, and note the plant and part of the plant on which the egg(s) is placed;
- presence of eggs and larvae on plants and feeding damage on plants.

c. Survey work

For those intending to make more detailed observations on island butterfly populations, a book on the techniques that can be applied is being prepared by one of us (TGS). Simple surveying techniques are described in Bennett & Humphries (1974). Another useful source is Dennis (1992), especially chapters 2 to 5 by T. G. Shreeve, K. Porter and M. S. Warren. On some islands, especially those which are nature reserves extensive monitoring of butterfly populations is already being carried out. For example, Skokholm and Skomer are sites in the ITE Butterfly Monitoring Scheme.

d. Publication of observations

A number of journals readily publish observations of Lepidoptera:

Butterfly Conservation News
The Bulletin of the Amateur Entomologists' Society
The Entomologist's Gazette
The Entomologist's Record and Journal of Variation.

The addresses for these journals are given in Colvin & Reavey (1993, *A Directory for Entomologists*). This also includes a great deal of other useful information. The most appropriate destination for island butterfly data is the computerized *Butterfly Net*, which is mapping records for the *Millennium Atlas Project*. Data can be sent to regional branch recorders or to the national recorder (Dr Jim Asher). The addresses for branch recorders and the national recorder can be obtained from:

The British Butterfly Conservation Society,
P.O. Box 222, Dedham,
Colchester, Essex CO7 6EY.

B. Butterfly records from an offshore island: the case example of Hilbre, Cheshire

Hilbre Island or Islands, as there are three of them (i.e., Hilbre, Middle Eye, Little Eye) off the Wirral peninsula, have one of the most complete sets of data on butterflies for any British offshore islands: the butterflies have been recorded continuously throughout the flight season for 11 years. Perhaps because of this the records for these islands demonstrate, more clearly than most others, the problems of determining the status of species on islands. Butterflies are monitored on the

Table 10. Species recorded and abundance measures for Hilbre between 1984 and 1994.

	1984	1985	1986	1987	1988	1989	1990	1991	1992	1993	1994	Host-plant	Incidence
T. sylvestris	0	0	0	0	0	0	0	+?	+	0	+	+	3
O. venata	0	0	1	1	0	0	0	0	0	0	0	+	2
C. croceus	0	0	0	0	0	1	0	0	0	0	0	+	1
P. brassicae	**	♦	*	+	**	♦♦	*	*	♦	**	*	+	11
R. rapae	**	♦	♦	**	♦♦	*	*	**	♦	**	**	+	11
P. napi	1	+	+	+	1	0	0	+	0	+	1	+	8
A. cardamines	0	0	0	0	0	0	1	0	+	0	0	–	2
L. phlaeas	**	*	+	*	+	*	0	+	+	+	+	+	10
P. icarus	**	**	0	0	+	*	**	+	*	+	*	+	9
V. atalanta	*	*	*	*	+	♦	*	**	**	*	**	+	11
C. cardui	0	*	0	0	*	+	+	*	*	0	+	+	7
A. urticae	♦♦	♦	**	**	♦♦	♦♦	♦	♦	♦	♦	**	+	11
I. io	*	+	0	0	0	0	+	+	*	♦	+	+	7
P. c-album	0	0	0	0	0	0	0	1	0	0	0	+	1
P. aegeria	0	0	1	1	+	1	1	1	1	1	*	+	9
L. megera	*	*	0	0	0	+	+	+	+	0	0	+	6
H. semele	0	1	0	0	1	*	0	0	0	1	+	+	5
P. tithonus	*	**	*	+	*	*	+	*	**	*	+	+	11
M. jurtina	+	*	**	*	**	♦	♦	*	*	+	0	+	10
C. pamphilus	0	1	1	0	0	0	0	0	0	0	0	+	2
No. of species	11	14	11	10	12	13	12	15	14	12	13		

0, no record; 1, one record; +, 2 to 9 records; *, 10 to 49 records; **, 50 to 99 records; ♦, 100 to 199 records; ♦♦, over 200 records; Hostplant; +, present, –, absent; Incidence, frequency of occurrence over 11 year period.

islands as part of the National Habitat Survey; some 90% of the records are believed to come from Hilbre itself. Until this year (1996) Hilbre's records have been made casually by staff usually at weekends, along with other aspects of wildlife, and the number of daily observations summed for each species to give a 'butterfly day total' (BDT). These figures, usefully annotated with dates of observations, are published in the annual *Hilbre Bird Observatory Reports* (Table 10).

These data provide a valuable documentation of changes in the fauna of the islands. However, the way the data are collected affects their interpretation. Observations have not been made as part of a fixed sampling design, for instance along a fixed transect as in the case of the Butterfly Monitoring Scheme (see Pollard & Yates, 1993; a fixed transect has been adopted on Hilbre for 1996). Thus, observations are not necessarily representative of habitats on the islands, nor can the variable 'transects' taken by observers be tied into quantitative estimates for the species. Strictly speaking, the numbers of records will not be comparable from year to year. There is also the issue of not being able to relate the records to a specific island. Nevertheless, the technique is almost certainly adequate to identify changes of the order of magnitude experienced during the decade on the islands as a whole. In the case of the Hilbre records, it is important to realize that the figures may give the impression that a species is more common on the islands than is actually the case. More than one 'transect' may be carried out on the same day by a different recorder. Thus, the same butterfly may be recorded by the same or different recorders on the same or different occasions. For example, though perhaps unlikely, the five records for *Pararge aegeria* made between 7–16 August could be of one individual. No systematic recording of early stages throughout the life cycle of species has been made, and there is therefore no real proof that species are established as breeding populations on the islands. The way in which the records are collected can lead to identification problems too. Observations of *Lasiommata maera* in 1990 and 1991 are obviously misidentifications for *L. megera*. The skippers seen in 1991 (probably *Thymelicus sylvestris*) are not accurately determined. The recorder for 1989 notes suspected misidentification problems for *Pieris brassicae* and *P. rapae*. Records for *Pieris napi* also seem to be grossly underscored compared to the number of vagrants observed over open land in Cheshire (Dennis, 1982a), never more than four observed in any year, and it is possible that some female *Anthocharis cardamines* are also lumped under *P. rapae*.

Nevertheless, several interesting features are evident in the records for the period from 1984 to 1994 that have been discussed at length in this work. First, although the numbers of species remains much the same (mean 12.4; standard error 0.45) the records indicate a turnover of species. As Hilbre consists of small tidal islands close to the mainland, this in itself is not surprising (see section I.3), but it is nice to see it confirmed. The contribution of species to the annual total, the species' richness of the islands at any time, is provided by the annual incidence of individual species. The probability of their presence on an island in any year is simply the frequency of their occurrence during the period of records, that is, assuming that conditions such as climate and habitats remain constant. However, conditions rarely do remain constant and it is this that makes the records of islands such as Hilbre interesting over time. Of course, data on a species' occurrence are based on the frequency of sightings or observations. This is dependent largely on how observations are made. The observations become increasingly reliable when a systematic recording scheme is in place. It once again emphasizes the importance of planning in survey design:

- the need to ensure that the scheme/survey is representative over space and time;
- that it is feasible, not over-ambitious, but can be undertaken by different observers to produce comparable results.

These issues are dealt with at length in Pollard & Yates (1993).

The cumulative total of species to 1994 is twenty-one, three of which are non-resident migrants. A further ten species are currently resident within about 50 km of the island (Emmet & Heath, 1989), and findings of this work suggests that these too may probably be observed on the islands given time and if observations continue at appropriate times of the year. It should be pointed out, though, that individuals of species that fail to find appropriate resources on a small island are unlikely to stay for any length of time on that island (see Shreeve, Dennis & Williams, 1995). Thus, observations will tend to be biased, as they are in gardens, to species which find suitable resources.

Some species, seen sparingly in odd years as singletons, are probably vagrant on Hilbre; examples are provided by *Thymelicus sylvestris*, *Ochlodes venata*, *Anthocharis cardamines*, *Polygonia c-album* and *Coenonympha pamphilus*. It is perhaps less likely that these species have bred successfully on the islands during the last decade than that they are casual visitors, but one cannot be certain as to the alternatives. Data on other species suggest that colonization and extinction have occurred. For example, *Polyommatus icarus* was absent in 1986 and 1987; this pattern is identical to records in terrestrial habitats over Cheshire (Dennis, pers. obs.). It is possible that it became extinct at the end of 1985 and recolonized the islands in 1988, but again the necessary data to back this supposition are missing. As 82 observations of the butterfly were made in 1984, it may seem unlikely that the butterfly subsequently became extinct. However, this figure of 82 needs to be placed in perspective. It is estimated by the recorder for the island that the 82 'individuals' observed may translate into as few as seven pairs of adults during the second brood in August 1984. Other species seem to have a similar pattern of colonization and extinction; for example, *Lycaena phlaeas*, *Pararge aegeria*, *Lasiommata megera* and *Maniola jurtina*. *Pararge aegeria* has produced an extraordinary series of single records for seven of the eleven years. All records have been made in September and October, with the exception of two years (i.e., May 1990 and August 1988). Then in 1994, 20 were seen on 22 September and 25 on 23 September. It seems more likely that these numbers are the product of colonization earlier in the same or a previous year than of vagrants entering directly from the Wirral. Whatever, it is not possible to dismiss the singletons in previous years as being vagrants. Multivoltine butterfly populations characteristically build up numbers throughout the season; the frequency of the butterfly may simply fail to exceed an observation threshold earlier in the year, but become numerous enough later on in the season to do so.

Occasionally, the records point to a probable source of vagrants and colonists. In the case of *Hipparchia semele* there are colonies at Red Rocks, an islet off the north-west Wirral coastline 1.6 km away, and on the dunes north of West Kirby. Knowledge of such potential sources, their proximity and size (i.e., area of colony and population size), is important in understanding the process of island colonization. The current analysis assumes that species within 50 km of the nearest mainland source to an island potentially form part of the faunal source. However, a species that is abundant and ubiquitous throughout the 50 km square is more likely to contribute to an island's fauna than a rare species well away from the shore in a more specialized habitat.

The Hilbre Islands records also illustrate how vulnerable species are, even those with the largest populations, on small islands. Large fluctuations are expected of migrant species which do not usually overwinter successfully in Britain (e.g., *Vanessa atalanta*, *Cynthia cardui*). But, similar fluctuations also affect species that very likely breed and overwinter on these small islands (e.g., *Inachis io*, *Maniola jurtina*). The population of *M. jurtina* shows signs of having built up to a peak from 1984 to 1990 and then to have declined to become extinct in 1994. The problem is again one of translating the 116 observations in 1990 into a measurement of effective population size. Such figures cannot usefully be guessed at; more detailed survey is necessary. A mark-release-recapture technique linked to a standard transect method for monitoring numbers (e.g., Thomas, 1983b) would be invaluable for furnishing precise data on population size, data that can be compared directly with changes in climate and habitat. Details on the latter require knowledge of the resources for species within habitats and mapping of these resources each year. For a small island, this is not as time consuming as may at first appear to be the case. It involves first mapping the area of the resources. Then, a measure of density or cover is required. This can be achieved by direct count or the systematic placement of quadrats (see Kershaw, 1964; Bennett & Humphries, 1974). To link populations of the species and their resources, that is to make sense of population fluctuations, even more detailed ecological survey is necessary. This would focus on the precise location of individuals, resource use, reproductive output and survival. Such detail lies outside the scope of this work. Techniques used in autecological surveys carried out on British butterflies can be gleaned from references in Dennis (1992). It is sufficient to indicate here that much can be learnt about butterfly populations on islands using only simple techniques. The data illustrated in this section for Hilbre, for all their shortcomings, clearly demonstrate this point. They are superior to information that we have for the majority of British and Irish offshore islands.